Improving the quality of apples

It is widely recognised that agriculture is a significant contributor to global warming and climate change. Agriculture needs to reduce its environmental impact and adapt to current climate change whilst still feeding a growing population, i.e. become more 'climate-smart'. Burleigh Dodds Science Publishing is playing its part in achieving this by bringing together key research on making the production of the world's most important crops and livestock products more sustainable.

Based on extensive research, our publications specifically target the challenge of climate-smart agriculture. In this way we are using 'smart publishing' to help achieve climate-smart agriculture.

Burleigh Dodds Science Publishing is an independent and innovative publisher delivering high quality customer-focused agricultural science content in both print and online formats for the academic and research communities. Our aim is to build a foundation of knowledge on which researchers can build to meet the challenge of climate-smart agriculture.

For more information about Burleigh Dodds Science Publishing simply call us on +44 (0) 1223 839365, email info@bdspublishing.com or alternatively please visit our website at www.bdspublishing.com.

Related titles:

Achieving sustainable cultivation of apples
Print (ISBN 978-1-78676-032-6); Online (ISBN 978-1-78676-034-0, 978-1-78676-035-7)

Achieving sustainable cultivation of temperate zone tree fruits and berries Volume 1: Physiology, genetics and cultivation
Print (ISBN 978-1-78676-208-5); Online (ISBN 978-1-78676-210-8, 978-1-78676-211-5)

Achieving sustainable cultivation of temperate zone tree fruits and berries Volume 2: Case studies
Print (ISBN 978-1-78676-212-2); Online (ISBN 978-1-78676-214-6, 978-1-78676-215-3)

Advances in postharvest management of horticultural produce
Print (ISBN 978-1-78676-288-7); Online (ISBN 978-1-78676-291-7, 978-1-78676-290-0)

Understanding and optimising the nutraceutical properties of fruit and vegetables
Print (ISBN 978-1-78676-850-6); Online (ISBN 978-1-78676-852-0, 978-1-78676-853-7)

Consumers and food: Understanding and shaping consumer behaviour
Print (ISBN 978-1-80146-354-6); Online (ISBN 978-1-80146-355-3, 978-1-80146-356-0)

Chapters are available individually from our online bookshop:
https://shop.bdspublishing.com

climate•SMART•publishing
IN AGRICULTURAL SCIENCE

BURLEIGH DODDS SERIES IN AGRICULTURAL SCIENCE
NUMBER 142

Improving the quality of apples

Edited by Professor Fabrizio Costa, University of Trento, Italy

burleigh dodds
SCIENCE PUBLISHING

Published by Burleigh Dodds Science Publishing Limited
82 High Street, Sawston, Cambridge CB22 3HJ, UK
www.bdspublishing.com

Burleigh Dodds Science Publishing, 1518 Walnut Street, Suite 900, Philadelphia, PA 19102-3406, USA

First published 2024 by Burleigh Dodds Science Publishing Limited

Library of Congress Control Number: 2023940753

British Library Cataloguing in Publication Data
A catalogue record for this book is available from the British Library

ISBN 978-1-80146-321-8 (print)
ISBN 978-1-80146-323-2 (PDF)
ISBN 978-1-80146-322-5 (ePub)
ISSN 2059-6936 (print)
ISSN 2059-6944 (online)

DOI: 10.19103/AS.2023.0127

Typeset by Deanta Global Publishing Services, Dublin, Ireland

Contents

Series list

Title	Series number
Achieving sustainable cultivation of maize - Vol 1 From improved varieties to local applications Edited by: Dr Dave Watson, CGIAR Maize Research Program Manager, CIMMYT, Mexico	001
Achieving sustainable cultivation of maize - Vol 2 Cultivation techniques, pest and disease control Edited by: Dr Dave Watson, CGIAR Maize Research Program Manager, CIMMYT, Mexico	002
Achieving sustainable cultivation of rice - Vol 1 Breeding for higher yield and quality Edited by: Prof. Takuji Sasaki, Tokyo University of Agriculture, Japan	003
Achieving sustainable cultivation of rice - Vol 2 Cultivation, pest and disease management Edited by: Prof. Takuji Sasaki, Tokyo University of Agriculture, Japan	004
Achieving sustainable cultivation of wheat - Vol 1 Breeding, quality traits, pests and diseases Edited by: Prof. Peter Langridge, The University of Adelaide, Australia	005
Achieving sustainable cultivation of wheat - Vol 2 Cultivation techniques Edited by: Prof. Peter Langridge, The University of Adelaide, Australia	006
Achieving sustainable cultivation of tomatoes Edited by: Dr Autar Mattoo, USDA-ARS, USA and Prof. Avtar Handa, Purdue University, USA	007
Achieving sustainable production of milk - Vol 1 Milk composition, genetics and breeding Edited by: Dr Nico van Belzen, International Dairy Federation (IDF), Belgium	008
Achieving sustainable production of milk - Vol 2 Safety, quality and sustainability Edited by: Dr Nico van Belzen, International Dairy Federation (IDF), Belgium	009
Achieving sustainable production of milk - Vol 3 Dairy herd management and welfare Edited by: Prof. John Webster, University of Bristol, UK	010
Ensuring safety and quality in the production of beef - Vol 1 Safety Edited by: Prof. Gary Acuff, Texas A&M University, USA and Prof. James Dickson, Iowa State University, USA	011
Ensuring safety and quality in the production of beef - Vol 2 Quality Edited by: Prof. Michael Dikeman, Kansas State University, USA	012
Achieving sustainable production of poultry meat - Vol 1 Safety, quality and sustainability Edited by: Prof. Steven C. Ricke, University of Arkansas, USA	013
Achieving sustainable production of poultry meat - Vol 2 Breeding and nutrition Edited by: Prof. Todd Applegate, University of Georgia, USA	014

Introduction

Apples are one of the most highly consumed fruits globally, with a worldwide global production estimated around 88 million tonnes each year. As a result of this popularity, consumers have extremely high expectations of the sensory quality of this fruit.

This volume provides a comprehensive review of the wealth of research on the processes which determine the key quality attributes of apples. The chapters are split into two parts: Part 1 chapters discuss topics such as consumer perception of apple quality, understanding texture development, the nutritional and nutraceutical properties of apples and the development of antioxidant nutraceutical compounds. Chapters in Part 2 review breeding and crop management to optimise quality, focusing specifically on aspects such as fruit quality improvement, pre-harvest apple fruit development and quality management and finally postharvest management of apple quality.

Part 1 Quality attributes

The first chapter of the book focuses on consumer perception of apple quality. Chapter 1 begins by providing an overview of the three key factors associated with consumer perception of apple quality: product-related, consumer-related and social environment-related factors. These points are then broken down into three individual sections. The chapter identifies the main product-related aspects, focusing specifically on intrinsic and extrinsic characteristics of apple. It then goes on to describe the consumer-factors by categorising them into psychological and physiological/biological influences. Social environment-related features are broken down into situational, socio-cultural and economic influences. The chapter also reviews future research considerations on consumer perception of apple quality.

Chapter 2 draws attention to the texture development of apples. It begins by first focusing on the anatomy and physiology of fruit texture traits. This is then followed by an evaluation of fruit texture parameters. The chapter goes on to discuss the influence of growing conditions, the effects of storage on fruit texture as well as the impact fungal diseases can have on fruit texture. A section on genetic determination of texture is also provided, followed by a review of the association between texture and aroma, as well as an overview of selection/breeding achievements.

Chapter 3 discusses the nutritional and nutraceutical properties of apples by reviewing the latest scientific studies on the health benefits of apples. The chapter begins by discussing the phytochemical composition of apples, then goes on to identify the nutraceuticals in apple products, such as apple puree, canned products, apple juice, apple sauces and other by-products such as pomace. A section on the role of apple nutraceutical compounds in promoting health and preventing disease is also included, focusing specifically on how consumption of these compounds has been correlated with positive outcomes for a variety of diseases, including cancer, diabetes, osteoporosis, cardiovascular disease, cognitive and lung disorders.

The final chapter of Part 1 focuses on the development of antioxidant nutraceutical compounds in apples. Chapter 4 begins by providing an overview of the nutritional composition of apples, drawing specific attention to macronutrients and micronutrients. The chapter moves on to discuss the various phytochemicals present in apples, providing individual sections on phenolic acids, flavonoids and antioxidants in essential oils. A section on the distribution of antioxidants in apples is also provided, which is followed by an overview of how oxidative damage to apples can impact their quality. Pre-harvest and postharvest changes in antioxidant content are also discussed and postharvest techniques preserving antioxidant content are reviewed. Finally, antioxidant bioavailability and the health benefits of antioxidants in apples are also discussed.

Part 2 Breeding and crop management to optimise quality

Part 2 opens with a chapter that examines breeding for fruit quality improvement in apples. Chapter 5 begins by providing an overview of apple quality trait phenotyping and its challenges, then examines how phenotypic selection can be used with and without DNA-based tools for apple quality breeding as well. The chapter also provides an overview of key fruit quality traits that are taken into consideration when breeding improved varieties, such as appearance, eating quality, and storability. A section on genome wide selection is also included in terms of how valuable a breeding approach it is in terms of apple quality traits.

Chapter 6 discusses advances in understanding pre-harvest apple fruit development. The chapter first provides an overview of key stages in apple fruit development, then goes on to review apple fruit growth models and precision orchard management. This is then followed by an overview of cell organisation and intercellular spaces in apple fruit growth. The chapter also highlights key environmental factors that can affect fruit growth, such as light, temperature, altitude and latitude. Agronomic factors affecting fruit growth such as crop load and thinning, irrigation and tree architecture are also described. A section on fruit development and pre-harvest fruit sensory quality is also included.

The subject of Chapter 7 is advances in pre-harvest management of apple quality. The chapter begins by highlighting key internal and external quality attributes that are directly influenced once apples are harvested as a basis for the rest of the chapter's discussion for how these attributes can be improved during the pre-harvest stage. The chapter identifies emerging issues affecting pre-harvest management of apple quality. It focuses specifically on environmental factors such as light and temperature, then highlights the importance of orchard management practices in improving apple quality.

The final chapter of the book discusses postharvest management of apple quality. Chapter 8 begins by identifying the key pre-storage factors that can influence apple quality during the postharvest period, such as pre-cooling, then examine how edible coatings can influence the nutritional and physicochemical properties of apple. Storage factors that affect apple quality are also described, specifically the use of plant growth regulators, ethylene and the various storage environments that can be used to maintain/improve apple quality. Other quality management technologies are also described, specifically the use of ozone, photocatalytic oxidation, cold plasma and the use of ultraviolet-C radiation and ultrasound technology. A section on potential postharvest disorders is also provided, which is then followed by an overview of the importance of using the right packaging and transportation methods in apple quality maintenance.

Part 1

Quality attributes

Chapter 1

Consumer perception of apple quality

Masoumeh Bejaei and Jennifer Arthur, Agriculture and Agri-Food Canada, Canada; and Margaret A. Cliff, The University of British Columbia, Canada

1 Introduction

Apples are the third most produced fruit in the world, next to bananas and watermelons (FAOSTAT, 2022). Apple production has increased steadily in the world over the last five decades, from 27 million tonnes in 1970 to more than 86 million tonnes in 2020 (FAOSTAT, 2022). The top ten apple producers in the world, in 2020, were China, USA, Turkey, Poland, India, Italy, Iran, Russian Federation, France and Chile, with a production of 40.50, 4.65, 4.30, 3.55, 2.73, 2.46, 2.21, 2.04, 1.62 and 1.62 million tonnes, respectively (FAOSTAT, 2022). The international trade value (i.e. export) of apples exceeded US$7.78 billion in 2019, and the top ten apple exporters in the world, in 2019, were China (16.01%), USA (12.35%), Italy (10.68%), European Union (10.19%), Chile (8.04%), New Zealand (7.29%), France (5.57%), South Africa (4.80%), Poland (4.72%) and the Netherlands (2.38%) (World Bank, 2021).

Apples are produced in different regions of the world and in both hemispheres (FAOSTAT, 2021), with the fruit having good storability and

http://dx.doi.org/10.19103/AS.2023.0127.01
Published by Burleigh Dodds Science Publishing Limited, 2024.

shelf life. As a result, apples are available for purchase year-round and are consumed in the majority of countries around the world. Although the average global supply of apples increased from 7.67 kg/capita/year to 8.03 kg/capita/year between 2009 and 2019, there was a huge variation in the per capita availability of apples among consumers from different countries. For example, fruit availability (both domestic production and import) was significantly higher than average in Albania, Democratic People's Republic of Korea, Turkey, Uzbekistan and Romania, with 88, 78, 77, 75 and 69 kg/capita/year, respectively, while 24 countries (e.g. Chad, Haiti and Sudan) had availability close to zero (FAOSTAT, 2021).

There are thousands of apple varieties in germplasm collections around the world, but only a few of them are actually grown and commercialized (Strohm, 2013). These collections consist of whole plants (fruit trees) in nurseries or orchards, cryoprotected dormant buds and seeds. Since apple seeds do not grow fruit that is true to type, apple varieties are propagated vegetatively by grafting buds or budwood.

Some varieties such as Gala, Red Delicious, Granny Smith, McIntosh and Golden Delicious were originally released as 'open varieties'. This means that the fruit trees or budwood could be purchased directly from the nursery, without a licensing agreement. In contrast, the majority of the new apple varieties are branded and released as 'club varieties' (see Subsection 3.2.3) that can only be grown and distributed by 'club' members (Brown and Maloney, 2009). Thus, the apple market has been transformed from a commodity market to one that also includes a segmented market - one that is even hypercompetitive (Musacchi and Serra, 2018).

Fruit quality definitions have changed for the industry and consumers (Brückner, 2022; Tijskens and Schouten, 2022). Consumers can have different perceptions and preferences of fruit quality that influence their purchase and consumption. These preferences are influenced by their psychological and physiological/biological state(s) and the sensory attributes of the fruit, as well as other social, economic and environmental factors (Nestle et al., 2009).

Multiple factors affect the attitudes, expectations, perceptions and preferences of consumers (Almli, 2012). Consumer perception and food preference studies are complex interdisciplinary areas of research involving biology (e.g. genetic factors and energy balance), physiology (e.g. gastrointestinal mechanisms), psychology (e.g. learning, personality traits, neophobia, expectations, motivation, decision-making, eating initiation and situational aspects), nutrition and food science (e.g. nutritional value and food chemistry), sensory and consumer science (e.g. sensory attributes and attitudes), market research (e.g. consumer attitudes, advertising and brands), economics (e.g. benefit, availability and price) and sociology (e.g. cultural traditions and function of food) (Köster, 2009).

Cardello (1995) discussed the complexity of food quality and defined it as: . . . *a consumer-based perceptual/evaluative construct that is relative to a person, place and time and that is subject to the same influences of context and expectations as are other perceptual/evaluative phenomena* (p. 163). Consequently, food choice, acceptance and quality concepts should be discussed, considering factors affecting products, consumers and their social environment (Meiselman, 2007).

The objectives of this chapter are (1) to present a schematic diagram to conceptualize the factors associated with the perception of apple quality, (2) to review the literature on the factors, and their interrelationships, that influence apple quality, and (3) to illustrate the dynamic apple market as consumer expectation and demand move toward global sustainability. The work has been assembled from a food technology and consumer science perspective and is not meant to be a comprehensive review of all disciplines that impact apple quality.

2 Schematic diagram illustrating the factors associated with consumer perception of apple quality

The schematic diagram shows a conceptualization of the main factors associated with consumer perception of apple quality (Fig. 1). It summarizes the perception of apple quality with product-related (🍎), consumer-related (👪) and social environment-related (👥) factors as well as some of the associated contributing variables and interrelationships. It was designed utilizing (1) a flowchart by Almli (2012), (2) consumer acceptance models by Khan and Hackler (1981) and Meiselman (2007), and (3) a schema by Mojet (copyright ATO 18-11-2001), as described in Köster (2009).

3 Influence of product-related factors on consumer perception of apple quality

The product-related factors (🍎) (Fig. 1) that are associated with the perception of apple quality are subdivided into two categories - intrinsic and extrinsic characteristics of the product. The intrinsic characteristics are those associated with the internal properties of the product, such as the sensory perceptions of taste, smell, texture and touch, whereas the extrinsic characteristics are those associated with the external properties of the product (e.g. packaging) that can be manipulated without fundamentally changing the product (Pramudya and Seo, 2019).

Intrinsic and extrinsic product characteristics are perceived directly or indirectly through human senses (i.e. sight, taste, smell, sound and touch), which influence fruit quality perceptions and consumer preferences (Ng et al., 2013; Endrizzi et al., 2015; Piqueras-Fiszman and Jaeger, 2015; Wang et al., 2019; Pramudya and Seo, 2019; Brückner, 2022).

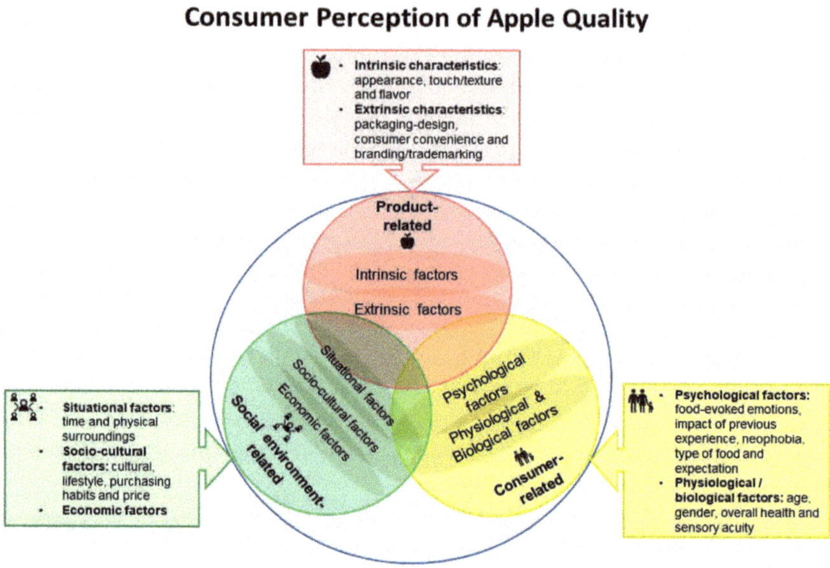

Figure 1 A schematic diagram summarizing the main factors associated with the perception of apple quality (product-related, consumer-related and social environment-related factors) as well as some of the contributing variables and some of their interrelationships.

3.1 Intrinsic product characteristics

The first influential factor for product-related apple quality is intrinsic attributes. The attributes of an apple can be categorized as appearance, texture/touch and flavor, from both external and internal assessments. Appearance includes the external characteristics such as fruit size, shape, symmetry, skin color, degree of coloration (over-color), degree of stripping (color variegation), degree of russeting (discolored patches), size of lenticels (small pore on apple cuticle), visual contrast of lenticels, lack of skin blemishes and skin shininess. While touch includes the external characteristics such as weight, smoothness, slipperiness and greasiness of the skin, texture includes the internal characteristics such as flesh hardness, crispiness, juiciness and skin toughness. Flavor is the combination of perceived sensations in the oral cavity (sweetness, tartness and bitterness) along with the aroma that is perceived via retro-nasal transfer (green apple, cooked apple, floral and perfume/spicy) (Slocombe et al., 2016).

Sensory and consumer tests, along with instrumental analysis (compositional and textural), can be useful in understanding and quantifying fruit characteristics. Such tools have been used for the selection and commercialization of new apple varieties (Hampson et al., 2000; Cliff et al., 2016; Teh et al., 2021).

3.1.1 Appearance attributes

The visual appearance of an apple strongly influences consumer purchase decisions (Kingston, 1992; Musacchi and Serra, 2018). Consumers consider the visual attributes of the fruit in evaluating the external quality, while the perception of the internal quality is based on texture, flavor and lack of flesh defects (Musacchi and Serra, 2018).

The outward appearance of the fruit is the main factor considered by consumers in making a purchase decision, while repeated purchases are influenced by the fruits' internal qualities (Barman et al., 2015). These decisions are influenced by previous experiences and by the selection and availability of varieties in the marketplace. With the majority of fruit in the marketplace being red- and bicolored apples, consumers are more familiar with them and may select them more often (Bejaei et al., 2020).

Skin color development is influenced by many factors including climatic conditions (e.g. temperature and light), production practices (e.g. trellising and pruning methods), maturity at harvest and ripeness after storage (Kingston, 1992; Lancaster and Dougall, 1992; Jing et al., 2020). During maturation of yellow apples, the chlorophyll (green color) in the skin and flesh is replaced by yellow pigments (Kingston, 1992). For red- and bicolored apples, the dominant skin color (over-color) is produced from anthocyanins from phenolic compounds, whereas the nondominant color (background color) is generated from the replacement of chlorophyll with yellow pigments (Lancaster and Dougall, 1992; Lancaster et al., 1994; Karagiannis et al., 2020). Dever et al. (1995) reported that apples of the same variety, with a higher percentage of red over-color, were sweeter and fruitier (Jonagold) or sweeter and with a higher pH (McIntosh). For bicolored apples, Dever et al. (1995) also demonstrated that the non-blush side of the apple was less sweet and crisper than the blush side of the fruit.

Flesh appearance and taste perceptions are correlated and serve as very important cues for shaping consumer perceptions and expectations. For example, consumers expect red apples to be sweet, green apples to be sour or tart and yellow apples to be sweet and/or have a softer flesh (Corollaro et al., 2014; Bowen and Grygorczyk, 2022).

Different types of skin disorders impact the fruit quality during production and storage. The skin disorders during production are russeting, scarf skin, sunburn and bitter pit, while the skin disorders during storage are superficial scald, senescent scald, lenticel blotch pit, skin necrosis and low-oxygen injury (Agriculture and Horticulture Development Board, 2021; BC Fruit Growers Association, 2021). Russeting is the occurrence of dark rough patches on the skin of the apple. This is influenced by environmental and genetic factors (Musacchi and Serra, 2018). While russeting negatively impacts the perception of fruit quality (Musacchi and Serra, 2018), Brown and Maloney (2003) point out that there may be market opportunities for some russeted varieties – as long as they have other desirable intrinsic traits. For example, consumers who have an interest in enhanced nutrition and future health benefits may accept russeted apples because they have a higher phenolic concentration and an increased antioxidant potential (Busatto et al., 2019).

Overall, the uniformity of size, shape and lack of any skin blemishes in apples are important to the majority of consumers (Kingston, 1992). The importance of external appearance has given rise to the selection of clonal variants called 'sports' (e.g. Gala and Fuji), with a higher percentage of red over-color (Wang et al., 2020).

The emphasis on visual appearance is so important that packinghouses wax apples to enhance their color and sheen to improve the perception of fruit quality. These factors are especially important for marketing loose (displayed or bulk) apples, since consumers can easily observe these visual defects. Skin color can actually intensify skin defects; for example, bruising is more evident in apples with light-colored rather than dark-colored skin. When minor visual defects are present on the skin of the fruit, consumers can be educated about the fact that the nutrition and food safety are unaltered, and the apples do not need to be discarded (Jaeger et al., 2018).

The majority of the consumers indicate that they like shiny apples. However, when they were informed that the apples were waxed at the packinghouse, they found this unacceptable and liked them less (Cliff et al., 2014a). The same consumers, however, changed their minds about the acceptability of the wax, when they were informed that it improves the storability of the apples by preventing dehydration (Cliff et al., 2014a).

External appearances of apples are important in quality assessments of fruits, for cross-cultural consumers and experts alike (Jaeger et al., 2018). The quality standards are very high for visual attributes, for which there is intense competition in the marketing of apple varieties. Jaeger et al. (2018) reported that fruit with minor defects were considered inferior compared to apples with no visual defects, and fruits with major defects were rejected by consumers – this pattern was very similar among consumers from different countries. Canadian consumers consider lighter colored fruit, which is grown in Canada, to be a sign of regional variation and natural production practices and did not downgrade the apple quality based on its skin color (Bejaei et al., unpublished data).

In addition to the impact that fruit genetics has on apple shape and size, environmental factors and agriculture practices (pruning, orchard practices and time of harvest) impact the visual characteristics of the fruit (Eccher and Noe, 1992; Charles et al., 2018). For example, Golden Delicious apples produced on steep terrain were more elongated and less russeted than those produced in valleys (Eccher and Noe, 1992).

The apple industry identifies the size of an apple as the number of fruit that are packed into a 40-lb box, which can range from 138 per box (i.e. extra-small apples) to 48 per box (i.e. extra-large apples). Most commercialized apple varieties are in the range of 113–88 apples per box; this corresponds to fruit that is 3″ (7.6 cm) to 3.25″ (8.3 cm) in diameter, respectively. However, most Canadians from all regions and ages prefer fruit to be ~3″ (7.4–7.6 cm) in diameter (Hampson et al., 2002). There are some exceptions if consumers are selecting apples for children's lunches (e.g. smaller and miniature) or gift-giving (e.g. larger and stenciled).

Fenko et al. (2010) reported that visual attributes were considered most frequently in the first stage of consumer-product interactions when consumers were less familiar with the products; however, as time passed, consumers shifted their focus to appreciating other sensations such as flavor (taste and smell) and touch. This suggests fruit retailers should customize their retailing strategy depending on the target consumers. They might select an apple variety with superior flavor characteristics for consumers with a higher consumption frequency, while a variety with superior visual attributes would likely be sufficient for the consumer segments with lower consumption frequency. Successful commercialization of a new variety in any market would require that the visual characteristics be at least comparable with the existing products in the marketplace – unless the apple has a particularly unique (and desirable), intrinsic or extrinsic characteristic(s).

Musacchi and Serra (2018) suggest that new apple varieties could be developed and introduced into the market to satisfy consumers with preferences for different skin colors (e.g. white- or black-skinned apples) and/or for different flesh colors (e.g. bright pink- to red-fleshed apples). This would generate novel apple markets (Subsection 6.2) and utilize branding and trademarking (Subsection 3.2.3).

3.1.2 Texture and touch attributes

Textural attributes of apples are intrinsic factors affecting consumers' perceptions and preferences of the fruit quality (Harker et al., 2003). Sensory textural cues for flesh firmness can be perceived by hand or mouth and can be used to access product quality attributes (Pramudya and Seo, 2019).

The majority of consumers prefer a crisp and juicy apple with low skin toughness, as identified by mouthfeel characteristics (Hampson et al., 2000; Harker et al., 2003; Cliff et al., 2016). Harker and Johnston (2008) explain that as fruit matures, apple texture softens and flavor develops. Indeed, the texture of apples is so important that the apple breeding programs have been using crispness and firmness as the selection criteria for improving the eating quality and storability of new apple varieties (Hampson and McKenzie, 2006; Hampson et al. 2009; Teh et al., 2020).

Published by Burleigh Dodds Science Publishing Limited, 2024.

In general, crisp and firm-textured apples are preferred - as long as the fruit is not too firm (Bowen and Grygorczyk, 2022). During storage, apple texture often softens and crispness declines (Costa et al., 2012). In some varieties, mealiness can develop after storage, especially if fruit had been harvested at advanced maturity (Harker and Hallett, 1992). Mealiness has been negatively correlated with crispness, firmness and juiciness (Bowen and Grygorczyk, 2022). As a result, the storage time and intended use of the apples may impact consumer choice, particularly if softer fruit is desired for cooking and baking. The delivery time to the market, temperature and humidity during transport and storage and retailer storage time also affect fruit quality.

There are many instruments that can measure the firmness of apples, using the force required to puncture peeled or unpeeled apples. These determinations are used for quality control at the packinghouse and are highly correlated with the sensory perception of firmness (Cliff and Bejaei, 2018; Bejaei et al., 2021; Bejaei, 2022); however, they do not simulate what consumers are experiencing when eating fresh fruit (Corollaro et al., 2014).

The surface texture of the fruit (an intrinsic factor) and the packaging (an extrinsic factor) both affect consumer perception, liking and purchase decision-making (Koutsimanis et al., 2012; Pramudya and Seo, 2019). Thus, the sense of touch is important in the selection of products (McCabe and Nowlis, 2003). This sense is closely related to emotions, and food-evoked emotions are better indicators of consumer acceptance than hedonic ratings (Dalenberg et al., 2014; Gutjar et al., 2015). More details on food-evoked emotions are described in Subsection 4.1.

Consumers and retailers evaluate the textural characteristics of fruit (especially firmness, roughness, coldness and slipperiness) by touching the product with their fingers and hands (Szczesniak and Bourne, 1969; Ranatunga et al., 2008).

Visual attributes are perceived as the first impression of the product, and the sense of touch confirms the visual impressions (Pramudya and Seo, 2019). As a result, new apple varieties should be offered first as display apples so that consumers can touch them at the point-of-sale. Later on, fruit could be offered as bagged apples after consumers have purchased and consumed the fruit, become familiar with the product and made the association with the name of the variety.

Multisensory integration of touch, sound (Schürmann et al., 2006) and tactile-taste interactions (Slocombe et al., 2016) also influence the perception of quality. The sourness of food is rated higher when it has a rough surface compared to a smooth surface (Slocombe et al., 2016). This knowledge could be applied in the selection of new varieties and designing new packages for bagged apples.

Cliff and Bejaei (2018) and Bejaei et al. (2021) have developed models that relate sensory and instrumental textural analyses. Such models allow for

Published by Burleigh Dodds Science Publishing Limited, 2024.

the rapid screening of a large number of apple varieties in an apple breeding program and circumvent the need for costly sensory panels. These models could also be used to evaluate fruit quality during production, transportation, storage and retail, thereby improving quality and minimizing waste at all stages of the supply chain.

3.1.3 Flavor attributes

Flavor is a multisensory sensation that occurs with stimulation of the taste buds, olfactory receptors and chemesthetic (trigeminal) receptors within the oral and nasal cavities (Slocombe et al., 2016). While sweetness is a sought-after flavor in most foods, bitterness is often disliked (Ventura and Worobey, 2013).

Flavor of apples plays a major role in consumers' preferences and perceptions (Hampson et al., 2000; Harker et al., 2003), and it is not limited to its sweetness and tartness (i.e. acidity). Volatile and nonvolatile compounds add complexity and richness to the flavor of apples that is appreciated by consumers (Aprea et al., 2012). Consumers are very much interested in eating flavorful apple varieties, and the chances of a repeat purchase are greater when an apple variety shows better flavor attributes (Harker et al., 2003). Aprea et al. (2012) identified 72 volatile compounds in different apple varieties and were able to successfully predict odor compounds responsible for nine flavor perceptions in apples.

Sweetness and tartness intensities have been evaluated in many apple sensory tests and differ among apple varieties (Hampson et al., 2000; Cliff et al., 2016; Cliff and Bejaei, 2018). Sweetness correlates positively with sorbitol and soluble solid concentrations (Aprea et al., 2017), until the upper threshold (i.e. terminal threshold) of sweetness has been achieved (Crisosto et al., 2006). A similar type of relationship was observed for titratable acidity and tartness. However, the perceptions of sweetness and tartness were dependent on each other (Crisosto et al., 2006) and vary simultaneously with consumer liking (Hampson et al., 2000; Harker et al., 2003).

Daillant-Spinnler et al. (1996) identified two segments of apple consumers in the British market. The first group preferred sweet and firm apples, while the second group preferred tart and juicy apples. Bejaei et al. (2020) suggested that different market segments may also occur for new and other varieties of apples that have unique flavors. This same research revealed that the majority of consumers purchased sweet apples, a smaller fraction purchased tart apples and the remaining consumers purchased apples with a balance of sweet and tart flavor (Bejaei et al., 2020). Their preferences were influenced by their cultural background (see Subsection 5.2).

Endrizzi et al. (2015) demonstrated that sweetness and texture acted together to influence the desirability of apples, with sweetness being the most influential factor. Consumer preference tests have also been used to identify liking and the factors affecting the purchase decision of new varieties (Cliff et al., 2014b, 2016). For example, Cliff and Toivonen (2017) reported that increasing harvest maturity resulted in an increase in the fruity flavor and sweetness perceptions and a decline in hardness, tartness and immature-green flavor perceptions in Ambrosia apples. These attributes (hardness, tartness and immature-green flavor) declined with increasing storage time up to 8 months (Cliff and Toivonen, 2017). Their work illustrated the importance of the use of sensory tests to track the influences of orchard practices prior to the commercialization of a new apple variety.

3.2 Extrinsic product characteristics

The second influential factor for product-related apple quality is extrinsic attributes. The extrinsic attributes can be altered without changing the product characteristics, for example by packaging (size and design), retail display, consumer convenience, food safety, production technology, organic status, genetic modification status and brand name (Jaeger, 2006; Becker et al., 2011; Piqueras-Fiszman and Spence, 2012; Bernard and Liu, 2017).

3.2.1 Packaging design and consumer convenience

Packaging contributes to the perceived quality of the product and affects sensory perceptions, product evaluations and consumer price expectations (Lefebvre et al., 2010; Becker et al., 2011). The store type and location may influence the demand for bagged or loose fruits (Cummins et al., 2009). Consumer demand for bagged apples has increased in large cities at major grocery stores but not in smaller cities at medium/small size grocery stores (Bejaei et al., unpublished data). The impact of packaging on consumer perception of fruit quality requires further investigation.

Lange et al. (2000) demonstrated that consumers make different purchase decisions for orange juice when they are exposed to just extrinsic information versus both extrinsic and intrinsic (flavor/taste) information. Cliff et al. (2010) also reported the importance of packaging in the perceived quality of fresh-cut apples. Indeed, there has been a growing interest in packaging studies to select the best packaging designs, packaging materials and packaging sizes (Pramudya and Seo, 2019).

Convenience (i.e. effortless consumption while saving time) and ease of preparation are important factors influencing consumer preferences in selecting whole fruit versus other fruits or snacks (Harker et al., 2003; Jaeger, 2006; Heng and House, 2018). With consumer demand for convenience, there is a growing interest in fresh-cut fruits and ready-to-eat foods (Baselice et al., 2017;

Pilone et al., 2017; Statista, 2021) in order to avoid processed foods (Montero-Vicente et al., 2019). The non-browning apple varieties are best for fresh-cut apple products because the apple slices will require little or no preservatives (Khanizadeh et al., 2007). For fruits, convenience can mean demand for year-round availability, maintaining quality without bruising and a clean easy-eating experience (Jaeger, 2006). New and other apple products that require sliced, juiced and dried apples may benefit from this trend, and a wider selection of varieties may be needed to fill in these niche markets (Subsection 6.2).

3.2.2 Production technology

Consumers from different cultures have varying perceptions and attitudes regarding different food production technologies, sustainability and locally sourced foods (Heng and House, 2018; Sánchez-Bravo et al., 2020; Jeong and Lee, 2021). In general, the education level of the respondents was positively associated with their knowledge and awareness about food sustainability issues, while consumers' age was negatively correlated with those concerns (Sánchez-Bravo et al., 2020; Jeong and Lee, 2021). Food safety is another important factor in food choice and food quality evaluation by consumers (Grunert et al., 2011).

Jaeger et al. (2004) had consumers weigh production technologies, depending on direct personal benefit. The production technologies were organic production, acceptance of genetically modified fruits, limited use of pesticides, hot/cold weather resistance and reduction in food waste. They found that when there was a direct personal benefit, less monetary compensation was necessary for consumers to accept the technology compared to an indirect general benefit to the environment (i.e. reduction of food waste) (Jaeger et al., 2004).

Organic, local production can influence quality perceptions. Bernard and Liu (2017) reported that US consumers perceived apples labeled organic and local to have higher sensory quality over unlabeled samples that were comparable.

3.2.3 Branding and trademarking

Apple varieties have traditionally been considered a commodity product, where varieties were readily available to the growers as 'open varieties'. This means that growers could purchase fruit trees directly from a nursery, without a licensing agreement. In previous generations, fruit quality was primarily defined using production-oriented criteria (yield, quantity, disease resistance, shelf life and production-season extension) (Brückner, 2022). When 'open varieties' were sold in bulk at low prices, there was little to distinguish them other than color or size (Canavari, 2018).

In an effort to increase returns, most new apple varieties are now released as 'club varieties' under a controlled management system. This means that growers must be a member or a grower-cooperative member of a 'club' in order to grow and sell a particular new variety (Brown and Maloney, 2009; Canavari, 2018). Member growers must pay royalties, contribute to marketing costs and must meet strict quality standards. In return, they receive a high revenue for their premium apples.

These new 'club varieties' are protected by a patent or plant variety rights. The new varieties are then marketed to consumers via trademarked brand names, which are required to be different than the varietal names. For example, the varietal name SPA493 from the Apple Breeding Program at Summerland Research and Development Centre (Hampson et al., 2013) is marketed to consumers under the Salish® apple brand (Summerland Varieties Corp., 2022). Trademarked brands help to differentiate apples in the marketplace and are the method by which consumers relate to them (Gontijo and Zhang, 2007) – with help from advertising and marketing that promotes the uniqueness, desirability and quality of the product (Hoegg and Alba, 2007).

The theoretical model of how consumers process information on the trademarked brand has been summarized by Vakratsas and Ambler (1999). A brand name is an extrinsic cue, which is easy to recognize compared to an intrinsic cue, such as sweetness (Olson and Jacoby, 1972; Jacoby et al., 1977; Zeithaml, 1988). Thus, a 'club variety' branding helps the consumers with purchase decisions, particularly for low-involvement consumers who are not familiar with the product category or are trying it for the first time (Zaichkowsky and Vipat, 1993). Branding serves the consumer and the retailer and prevents or avoids an apple variety from falling into the commodity category.

Consumers consider new, managed varieties as a separate market – different from traditional apples (Rickard et al., 2013). Keller (2003) and Keller and Swaminathan (2020) identified that brands helped consumers identify the product origin, assign responsibilities to manufacturers, reduce risks, simplify choices, reduce the costs of searching for a product and build a stronger relationship between consumers and manufacturers and are a symbol of quality.

4 Influence of consumer-related factors on consumer perception of apple quality

The consumer-related factors () (Fig. 1) associated with the perception of apple quality are subdivided into two categories – psychological and physiological/biological factors. The psychological factors include food-evoked emotions, impact of previous experience, neophobia, types of food and food expectation, whereas the physiological/biological factors include age, gender, health and sensory acuity. Some of these factors were mentioned briefly in previous sections when discussing their interactions with intrinsic and extrinsic product characteristics.

4.1 Psychological factors

The development of food preferences starts before or immediately following birth (Mennella et al., 2001; Ventura and Worobey, 2013). Early food exposure, food preferences, neophobia and unconscious learning all impact food choice throughout one's life span (Birch, 1990; Nicklaus et al., 2005). Food selection for eating is a learned behavior (Köster, 2009), with food neophobia (fear of unfamiliar foods) sometimes emerging when a child is 2 years old (Birch, 1990).

Food-evoked emotions affect the food choice and food quality perception of consumers (Cardello et al., 2012; Kenney and Adhikari, 2016). There has been much interest in studying emotional responses to food because of its importance in shaping consumer preference and perception. Cardello et al. (2012) demonstrated that the magnitude of emotional responses of consumers is dependent on the type of food and familiarity with the product.

Online surveys have been used to evaluate consumer food-based conscious emotions. Ares et al. (2015) found that the food-based 'well-being' responses were highly correlated with pleasure, physical health and emotional aspects. Sulmont-Rossé et al. (2019) found that food-related 'feeling good' responses varied not only with the sensory characteristics but also with knowledge of the nutritional value and the future health benefits of the foods.

The study of unconscious emotions has also been evaluated. Köster (2009) states that pleasurable feelings shape the consumers' unconscious sensory expectations that are associated with food quality. The author points out that implicit assessments of preference should accompany explicit measurements since consumers may not be able to conceptualize their reasons for liking a product. Köster (2009) suggests using past behavior and indirect (observational) methods for assessing consumer attitudes and preferences because it will provide more reliable data on food choice compared to asking the consumer directly. Indeed, Bejaei et al. (2020) used this approach to study past and current consumer apple purchases, as described in Subsections 5.2 and 5.3.

Raudenbush et al. (1998) studied individuals who avoided (neophobics) or liked (neophilics) new foods. While the individuals' ratings for the pleasantness of the foods were similar between the two groups, neophobics were very much less willing to try new and novel foods. In an effort to understand neophobia, Tuorila et al. (1994) determined that the best way to convince neophobics that the food 'tastes good' was to have them 'take a bite'. This behavior may present a challenge for the apple industry when trying to introduce new apple products and varietals.

Köster (2009) indicates that the impact of previous food experience and memory also play an active and important role in the perception of food sensory characteristics. Even though sensory perception may change over time (e.g. due to aging), some older consumers may maintain their choices, due to their reliance on memory and previous food experience (Mojet et al., 2005). In addition, cultural differences influence the perception of apple varieties (Cliff et al., 1999, 2014b), as discussed in Subsection 5.2.

Collinsworth et al. (2014) explored relationships between texture and emotion using images (textural) and words. They found that the impact of the images, on the emotional response to food, was greater in females than males. Thus, the emotional responses to the food can be used to predict consumer liking and preference (Samant et al., 2017) as well as assist with identifying food choice behavior (Gutjar et al., 2015). Consumers' emotional responses, to images of apples with different characteristics, could be explored using this method.

Although the emotional factors are complex, they are influenced by other factors such as physiological/biological, situational and sociocultural, as discussed in Subsections 4.2, 5.1 and 5.2, respectively. In fact, the majority of the factors were studied individually. Some authors, however, were able to study a few factors simultaneously using conjoint analysis (Deliza et al., 2010; Endrizzi et al., 2015; Almli and Næs, 2018) and using experimental markets (Lusk, 2003; Jaeger et al., 2004; Lund et al., 2006). This area of research is particularly difficult due to the large number of variables involved (Meiselman, 2007; Köster, 2009). Lüscher et al. (2013) and Ossadnik et al. (2016) have used a pairwise-comparison-weighting method to look at multiple variables simultaneously. Future work could consider expanding the use of this methodology for capturing the influence of product-related (Section 3), consumer-related (Section 4) and social environment-related (Section 5) factors.

4.2 Physiological and biological factors

Flavor perception starts with the stimulation of taste receptors (taste buds) in the mouth, olfactory receptors in the nose and trigeminal receptors that are located throughout the oral and nasal cavities. While taste refers to the perception of five basic modalities (sweet, sour, salty, bitter and umami) in the mouth, flavor refers to the combination of taste, smell, texture (mouthfeel) and irritation that occur throughout the oral and nasal cavities. These terms, taste and flavor, are often used interchangeably in colloquial conversation.

Sensory perceptions vary with genetics and start before or at birth (Feeney et al., 2011; Guido et al., 2016; Barragán et al., 2018). An et al. (2018) found that physiological changes, with age, affect mastication, digestion, gastrointestinal physiology, the quantity and quality of absorbed nutrients and the diversity of intestinal microbiota.

Many studies have shown that flavor perception declines with age (Murphy, 1993; Mojet et al., 2003, 2005; Kennedy et al., 2010; Barragán et al., 2018), but its decline is not identical for different taste sensations (i.e. sweet, sour, salty, bitter and umami), different types of foods or different measures of olfactory acuity (i.e. odor detection, odor identification and odor intensity) (Pangborn and Pecore, 1982; Murphy, 1993; Barber, 1997; Mojet et al., 2003, 2005; Guido et al., 2016; Barragán et al., 2018).

Murphy (1993) and Mojet et al. (2003) reported that the sense of taste diminishes more with age than the sense of smell. Some loss in olfactory performance can be 'compensated for', in older adults, by their enhanced experience and/or familiarity with the food (Barber, 1997; Murphy, 1993).

The literature, however, is inconclusive regarding the relationship between taste preferences (i.e. hedonic ratings) and taste acuity (taste detection thresholds or supra-threshold intensity). While Pangborn and Pecore (1982) found relationships between food preference and taste intensity, other researchers did not (Chauhan and Hawrysh, 1988; Guido et al., 2016). Furthermore, Mojet et al. (2005) reported that individuals of different ages preferred almost identical concentrations for four of the five basic taste sensations (sour, salty, bitter and umami) – as long as the individuals were still able to smell. This inconsistency in the literature suggests that flavor preferences are complex, with much to be yet understood about their development and evolution over time.

Similarly, vision and hearing are also negatively impacted by age (Aydelott et al., 2010; Andersen, 2012). These sensory modalities can have a major impact on taste perception, health and well-being of older adults. The sensory limitations in older adults can be partially rectified with technological aids (glasses and hearing aids), but no such aid is available for flavor perception. In addition, other confounding factors may influence sensory perception in older adults, including ethnic background, residency, prescription drugs, smoking, head injury, chemical exposure, overall health (Barber, 1997; Guido et al., 2016) and disease (McMahon et al., 2014). McMahon et al. (2014) found that individuals with chronic kidney disease had altered perceptions of sour, salty and umami independent of patients' age and gender.

The literature has also mixed results regarding the influence of gender on sensory perceptions (Chauhan and Hawrysh, 1988; Barber, 1997; Barragán et al., 2018). Chauhan and Hawrysh (1988) reported females to have higher intensity scores for the different taste sensations, but this has not been corroborated by other researchers (Mojet et al., 2001; Guido et al., 2016). Even though 6-*n*-propyl-thiouracil (PTU) status is commonly utilized as an index of 'taste sensitivity' by many researchers, Dinnella et al. (2018) demonstrated that the relationship between PTU status and taste bud density was not as straightforward as once believed.

Females have also been shown to vary in their taste perception (particularly sweetness) with body mass index (BMI) (Feeney et al., 2011; Guido et al., 2016). Proserpio et al. (2017) found that women with higher BMI perceived sweetness and vanilla flavor more intensely than men with similar BMI. In other research, Rosa et al. (2022) found that women with higher BMI perceived odor more intensely but perceived salty and bitter less intensely. They concluded that body weight and gender played an important role in sensory perception.

Chao et al. (2017) identified that females, who were fasting, were more responsive to visual food cues, as quantified by higher activity of the frontal, limbic, striatal and fusiform gyrus parts of the brain.

Clearly, flavor perception is affected by many underlying physiological and biological factors and individual differences based on age, gender, overall health, genetics, sensory acuity and oro-gastrointestinal physiology. These in turn influence food preferences and food choices and are influenced by social environment-related factors, as discussed in Section 5.

5 Influence of social environment-related factors on consumer perception of apple quality

The social environment-related factors () (Fig. 1) that are associated with the perception of apple quality are subdivided into three categories: situational factors (e.g. time and physical surroundings), the sociocultural factors (e.g. cultural, lifestyle, purchasing habits and price) and economic factors (e.g. price to value and online shopping).

5.1 Situational factors

Consumers' food perceptions and preferences are highly dependent on the context and situation in which the food is consumed (i.e. consumption time, location, hunger state and social settings) (Jaeger, 2006; Hein et al., 2012; Piqueras-Fiszman and Jaeger, 2015; Olegario et al., 2021). The same individual may appreciate the same food differently in different situations (Hein et al., 2012). The context and situation can evoke different emotions (Jaeger, 2006; Romeo-Arroyo et al., 2022) and can influence the use of the product, its convenience and eating behavior. Eating an apple was considered less convenient in the presence of others, while standing and in the evening (7:00 pm) compared to the afternoon (3:00 pm) - as apples are considered a snack item in western cultures (Jaeger, 2006).

Hein et al. (2012) evaluated consumers' liking for apple and blackcurrant juices under different consumption contexts. They used written scenarios to clarify the expectations and conditions for all respondents and found that the more appropriate the consumption context, the stronger the positive emotions and hedonic ratings.

Indeed, food-evoked emotional responses were context dependent (Piqueras-Fiszman and Jaeger, 2015; Kenney and Adhikari, 2016). Gutjar et al. (2015) demonstrated that, for breakfast drinks and desserts, the emotions evoked by intrinsic (sensory) and extrinsic (packaging) characteristics were different. When intrinsic information was made available (via tasting), the participants responded differently compared to when only extrinsic information was available.

5.2 Sociocultural factors

Some of the factors shaping the attitudes of consumers are culture, familiarity with food, local availability of the products, values, lifestyle, purchasing habits and sociodemographic characteristics (Almli, 2012; Heng and House, 2018; Jeong and Lee, 2021).

Galmarini et al. (2013) reported differences in fruit quality expectations in a cross-cultural study with consumers from Argentina and France. Textural attributes were determined to be equally important for all respondents, with consumers agreeing on the definition of a 'good quality apple', using the terms juicy, color, crunchy, sweet, tasty, firm, texture and fresh characteristics - in their own languages. However, Argentinian consumers, with a lower apple consumption frequency, first considered visual attributes (especially skin color) and then textural and flavor attributes (in that order) for describing apple quality. In contrast, French consumers with higher apple consumption frequency considered first flavor and then visual attributes for describing the quality of apples (Galmarini et al., 2013) - and were willing to pay more (Montero-Vicente et al., 2019). Interestingly, the French consumers who consumed more apples and were more familiar with apple varieties used more specific and fewer words to describe the flavor attributes than the Argentinian consumers (Galmarini et al., 2013).

Cliff et al. (1999) reported differences in the perceived apple flavor and texture attributes by consumers from two Canadian provinces (i.e. British Columbia (BC) and Nova Scotia (NS)) - located 4500 km apart. The NS consumers rated the texture and flavor significantly higher than BC consumers, possibly because they were accustomed to smaller, lower-quality apples in their marketplace at the time. As for the visual preferences, consumers from both provinces preferred varieties that were commonly grown in their own region. This not only illustrates regional differences in consumer preference but the importance of familiarity with locally sourced apples.

Research by Jaeger et al. (1998) did not find differences in apple variety preferences between British and Danish consumers for three apple varieties; however, this lack of difference may have been related to the fact that the majority of their consumers had European backgrounds.

Cliff et al. (2014b) also reported differences in apple variety preferences of consumers based on their ethnicity (i.e. ancestral origin) and preferred apple type (i.e. sweet or tart). In their study, 88% of participants of Asian ethnicity identified themselves as sweet-apple eaters. In contrast, 55% of participants of European ethnicity identified themselves as sweet-apple eaters and the remainder (45%) identified themselves as tart-apple eaters. In short, participants' hedonic scores for the different types of apples were consistent with, and matched, their stated preferred apple type. This demonstrates the importance of considering the target market when introducing new varieties.

Bejaei et al. (2020) also matched fruit purchases of participants, with a list of the intrinsic characteristics (for each variety), to indirectly identify the consumers' taste preferences. They also reported that 80.0% of the participants with Asian and Southeast Asian backgrounds purchased sweet apples, while 57.9% and 25.0% of participants of European background chose sweet and sweet-tart apples, respectively.

Fruits can also have different social functions in different cultures (Heng and House, 2018). For instance, fruits often carry a symbolic meaning in Chinese culture and are served as part of a formal dinner (Ma, 2015). However, there can be challenges with the introduction of new varieties into different cultures (Andani et al., 2001) and regions (Cliff et al., 1999) because consumers tend to prefer familiar varieties. Convincing consumers to try new and unique varieties often requires additional marketing efforts and/or consumer education.

Heng and House (2018) reported that the consumption frequency of fruits can be different in different countries. For example, French consumers eat more fruit than consumers in the USA (Tamers et al., 2009).

Montero-Vicente et al. (2019) conducted a study in Spain, where 90% of the population eat fruit daily. Based on the consumers' perception of health and nutritional benefits, the authors identified four segments of the market – 'total indifference', 'little time to cook, concerned about nutrition…', 'cooks, preference for natural products' and 'unconcerned' – representing 4.0%, 26.4%, 40.0% and 29.4%, respectively. This suggests that introduction of a fruit with new enhanced health or nutritional benefits (e.g. higher antioxidant content) may not necessarily result in widespread demand for the product (Endrizzi et al., 2015).

Heng and House (2018) evaluated the drivers of fruit consumption and consumer purchase behavior in eight countries in North America, Europe and East Asia. They were able to identify three apple market segments: 'low-frequency', 'common fruit' and 'high-frequency' consumer groups who were different in their socioeconomic classifications. The 'low-frequency' consumer segment consisted of consumers who were younger, single, unhealthy (self-reported), not active and focused on apple price, while the 'common fruit' and 'high-frequency' consumer segments consisted of individuals who were older, married, healthy (self-reported), physically active and focused on the fruits' attributes (freshness, nutritional value, origin and in-season) with very little attention to apple price.

5.3 Economic factors

Price is another factor affecting consumer food choice. In a commodity market, the purchase of apples is dependent on the price (Harker et al., 2003; Jaeger, 2006). Richards and Patterson (2000) found that consumers who purchase apples out of habit are more affected by the price of apples.

Under the 'club variety' system, apple varieties with a specific intrinsic or extrinsic characteristic are often branded to capitalize on their unique trait(s). This branding allows the quantity and quality of the supplied fruit to be managed to achieve a premium price (Canavari, 2018). Consumer price sensitivity and willingness to pay (WTP) depend on many factors (Heng and House, 2018). For consumers whose food costs represent only a fraction of their income, a premium price paid for a fruit may be easily accepted. In contrast, for other consumers whose food costs are a substantial portion their income, they may not be willing or able to pay a premium for these products. These price-sensitive consumers have a different concept of price to value and may also have a different availability of fruit in their local (neighborhood) grocery store. Some consumers, who are not price sensitive, may seek higher prices as an indication of a higher quality product or as a status symbol (Lichtenstein et al., 1993; Jaeger, 2006; Ma, 2015; Heng and House, 2018).

When conducting consumer research, Jaeger (2006) cautions the use of WTP questions on consumer ballots because customers may change their purchase behavior when shopping. This means that consumer research is more effectively conducted by testing the products in experimental markets or studying consumers' past purchase behavior (Jaeger, 2006). As such, Bejaei et al. (2020) studied the actual purchase behavior of consumers in the retail market and at a special event (apple festival) market and classified apple consumers into different market segments, based on their purchase decisions for the various apple varieties. This indicates there is a delicate balance between WTP and situational decision-making.

Most recently, there has been a change in consumers' shopping behavior due to the COVID-19 pandemic (2019–2022). Many consumers are now shopping online for food and other products (Alaimo et al., 2020). Meißner et al. (2020) indicated that consumers shopping online choose a larger variety of products and were less sensitive to price. Although this study did not involve fruit, the apple industry needs to know how this new shopping behavior affects the purchase of apples.

6 Future research considerations on consumer perception of apple quality

Consumer perception of apple quality is not static, as reviewed in Sections 3-5 and summarized in Fig. 1. Staying attuned to changing consumer behavior, using the latest technologies, will provide valuable insight into the dynamic apple market.

6.1 Virtual reality technology

Virtual reality (VR) is a new frontier and useful tool in sensory science and consumer situational research, to create digital environments to understand consumer behavior (Jerald, 2015; Javornik, 2016; Kong et al., 2020). Javornik (2016) found that virtual food, or food-related cues, and digital environments can successfully induce food cravings. VR could also be used to gather information, in highly realistic contextual environments, on consumers' preferences at point of sale and point of consumption. Van Herpen et al. (2016) demonstrated that consumer behavior in a VR store was comparable with behavior in a physical store. Wang et al. (2021) suggested that online supermarkets could add a new feature to suggest healthy dish selections and to promote healthy eating habits, thereby removing the link between healthy food and 'boring' food.

In 2009, the European Union relaxed regulations regarding irregular/blemished fruit in the marketplace in order to reduce food waste (Lombart et al., 2019). Lombart et al. (2019) used VR to examine consumer perception and purchasing behavior for abnormally shaped fruits and vegetables. They found their acceptance, in a virtual grocery store, was related to the degree of abnormality. The apple industry could benefit from understanding the acceptance of irregular/blemished fruit in the marketplace.

Indeed, VR technology is a useful tool for studying food choices and understanding the dynamic apple market. Through game-play and repeat-exposes to new apple products, VR could be used to influence consumers to make sustainable choices (Wang et al., 2021). As such, this could translate into increased acceptance and liking of new sustainable products.

6.2 The dynamic apple market requires multidisciplinary and interdisciplinary research

As the agricultural regions around the world are affected by climate change and emergencies (hurricanes, flooding, temperature extremes and pandemics), the production, distribution and quality of apples will change. To maintain high-quality fruit, producers will need to change their agricultural practices and researchers will need to expand their scope to incorporate multidisciplinary and interdisciplinary fields.

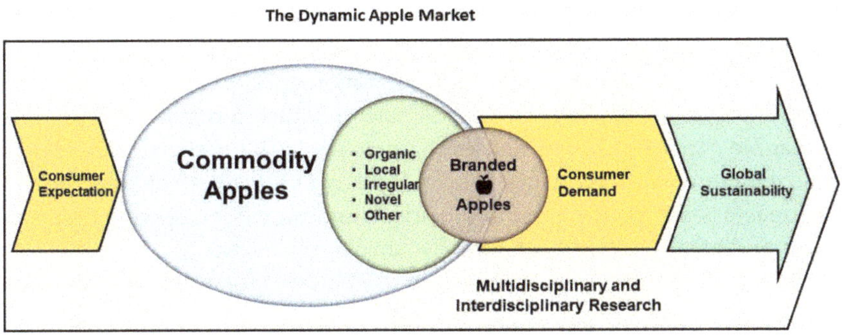

Figure 2 A schematic diagram of the dynamic apple market, showing the commodity and specialty apple markets with ellipses and their interrelationship with the branded apple market.

The dynamic apple market is represented in Fig. 2. The larger ellipse in Fig. 2 represents the commodity market, with the smaller ellipse representing specialty markets: organic, local, irregular, novel (e.g. miniature, stenciled, heirloom, white-skinned, early-harvest and hot-climate apples) and other apple products (e.g. sliced, juiced, dried, non-browning and astringent apples). The circle identifying the branded apple market overlaps with all ellipses, representing the interaction of the various apple markets. Branded apples are known for their high-quality standards; however, substandard branded fruit should still be marketed, without the brand name (e.g. with the variety name) in order to reduce food waste. All markets are driven by consumer expectation and consumer demand and moving toward global sustainability.

Overall global sustainability is imperative to maintain water quality, lower chemical residues on food, maintain higher soil quality, minimize soil degradation, encourage biodiversity and reduce reliance on nonrenewable resources (Reganold et al., 2001). Jin et al. (2022) identify that environmental and sustainability must be considered, simultaneously with economic and social issues, since production of food and the livelihoods of producers are at stake.

7 Overview of consumer perception of apple quality

Apples are hugely important on the international stage and will continue to be so in the future. This chapter reviews apple quality and the many factors that influence consumer preference. Intrinsic apple characteristics such as appearance, texture, flavor and size, as well as extrinsic characteristics such as branding, convenience, packaging and marketing, all impact the perception of apple quality. These perceptions are also affected by psychological, physiological and biological factors, in the context of the social environment-related (situational, sociocultural and economic) factors. There are other influences at the local, regional, national and international levels that can affect consumer preference. Currently, consumers have high expectations for apple quality - with visual appearance being paramount. As apple production and supply chain issues are impacted by climate change, consumers will prioritize other apple characteristics as well as accept irregular/blemished fruit, as part of their commitment to global sustainability. In this chapter, two schematic diagrams were developed and presented to provide a framework for communicating and understanding apple quality and the dynamic apple market.

8 Where to look for further information

- Almli, V. L. and Næs, T. (2018), Conjoint analysis in sensory and consumer science: Principles, applications, and future perspectives. In: Ares, G. (Ed.), *Methods in Consumer Research, New Approaches to Classic Methods* (vol. 1), Woodhead Publishing, Cambridge, pp. 485–529.
- Harker, F. R. and Johnston, J. W. (2008), Importance of texture in fruit and its interaction with flavour. In: Brückner, B. and Wyllie, S. G. (Eds.), *Fruit and Vegetable Flavour: Recent Advances and Future Prospects*, Woodhead Publishing, Cambridge, pp. 132–149.
- Keller, K. L. and Swaminathan, V. (2020), *Strategic Brand Management: Building, Measuring, and Managing Brand Equity* (5th edn.), Pearson, Harlow UK, 624 p.
- Meiselman, H. L. (2007), Integrating consumer responses to food products. In: MacFie, H. J. H. (Ed.), *Consumer Led Food Product Development*, Woodhead Publishing, Cambridge, pp. 3-33.

9 References

Agriculture and Horticulture Development Board (2021). Apple best practice guide, Available at: https://apples.ahdb.org.uk/; accessed April 15th, 2022.

Alaimo, L. S., Fiore, M. and Galati, A. (2020). How the Covid-19 pandemic is changing online food shopping human behaviour in Italy, *Sustainability* 12(22), 9594.

Almli, V. L. (2012). Consumer acceptance of innovations in traditional food. Attitudes, expectations and perception, Doctoral Dissertation, Norwegian University of Life Sciences.

Almli, V. L. and Næs, T. (2018). Conjoint analysis in sensory and consumer science: principles, applications, and future perspectives. In: Ares, G. (Ed.), *Methods in Consumer Research: New Approaches to Classic Methods* (vol. 1), Woodhead Publishing, Cambridge, pp. 485–529.

An, R., Wilms, E., Masclee, A. A. M., Smidt, H., Zoetendal, E. G. and Jonkers, D. (2018). Age-dependent changes in GI physiology and microbiota: time to reconsider?, *Gut* 67(12), 2213–2222.

Andani, Z., Jaeger, S. R., Wakeling, I. and MacFie, H. J. H. (2001). Mealiness in apples: towards a multilingual consumer vocabulary, *J. Food Sci.* 66(6), 872–879.

Andersen, G. J. (2012). Aging and vision: changes in function and performance from optics to perception, *Wiley Interdiscip. Rev. Cogn. Sci.* 3(3), 403–410.

Aprea, E., Charles, M., Endrizzi, I., Laura Corollaro, M., Betta, E., Biasioli, F. and Gasperi, F. (2017). Sweet taste in apple: the role of sorbitol, individual sugars, organic acids and volatile compounds, *Sci. Rep.* 7(1), 44950.

Aprea, E., Corollaro, M. L., Betta, E., Endrizzi, I., Demattè, M. L., Biasioli, F. and Gasperi, F. (2012). Sensory and instrumental profiling of 18 apple cultivars to investigate the relation between perceived quality and odour and flavour, *Food Res. Int.* 49(2), 677–686.

Ares, G., de Saldamando, L., Giménez, A., Claret, A., Cunha, L. M., Guerrero, L., de Moura, A. P., Oliveira, D. C. R., Symoneaux, R. and Deliza, R. (2015). Consumers associations with wellbeing in a food-related context: a cross-cultural study, *Food Qual. Prefer.* 40, 304–315.

Aydelott, J., Leech, R. and Crinion, J. (2010). Normal adult aging and the contextual influences affecting speech and meaningful sound perception, *Trends Amplif.* 14(4), 218–232.

Barber, C. E. (1997). Olfactory acuity as a function of age and gender: a comparison of African and American samples, *Int. J. Aging Hum. Dev.* 44(4), 317–334.

Barman, K., Ahmad, M. S. and Siddiqui, M. W. (2015). Factors affecting the quality of fruits and vegetables: recent understandings. In: Siddiqui, M. W. (Ed.), *Postharvest Biology and Technology of Horticultural Crops: Principles and Practices for Quality Maintenance* (1st edn.), Apple Academic Press, New York, pp. 1–50.

Barragán, R., Coltell, O., Portolés, O., Asensio, E. M., Sorlí, J. V., Ortega-Azorín, C., González, J. I., Sáiz, C., Fernández-Carrión, R., Ordovas, J. M. and Corella, D. (2018). Bitter, sweet, salty, sour and umami taste perception decreases with age: sex-specific analysis, modulation by genetic variants and taste-preference associations in 18 to 80 year-old subjects, *Nutrients* 10(10), 1539.

Baselice, A., Colantuoni, F., Lass, D. A., Nardone, G. and Stasi, A. (2017). Trends in EU consumers' attitude towards fresh-cut fruit and vegetables, *Food Qual. Prefer.* 59, 87–96.

BC Fruit Growers Association (2021). BC tree fruit production guide – your one stop guide for managing your crops, Available at: https://www.bctfpg.ca/; accessed April 15th, 2022.

Becker, L., van Rompay, T. J. L., Schifferstein, H. N. J. and Galetzka, M. (2011). Tough package, strong taste: the influence of packaging design on taste impressions and product evaluations, *Food Qual. Prefer.* 22(1), 17–23.

Bejaei, M. (2022). Converting apple textural parameters obtained from penetrometers and their relationships with sensory attributes, *Horticulturae* 8(3), 269.

Bejaei, M., Cliff, M. A. and Singh, A. (2020). Multiple correspondence and hierarchical cluster analyses for the profiling of fresh apple customers using data from two marketplaces, *Foods* 9(7), 873.

Bejaei, M., Stanich, K. and Cliff, M. A. (2021). Modelling and classification of apple textural attributes using sensory, instrumental and compositional analyses, *Foods* 10(2), 384.

Bernard, J. C. and Liu, Y. (2017). Are beliefs stronger than taste? A field experiment on organic and local apples, *Food Qual. Prefer.* 61, 55–62.

Birch, L. L. (1990). The control of food intake by young children: the role of learning. In: Capaldi, E. D. and Powley, T. L. (Eds.), *Taste, Experience and Feeding*. American Psychological Association, Washington DC, pp. 116–135.

Bowen, A. and Grygorczyk, A. (2022). Consumer eating habits and perceptions of fresh produce quality. In: Florkowski, W., Banks, N., Shewfelt, R. and Prussia, S. (Eds.), *Postharvest Handling: A Systems Approach* (4th edn.), Academic Press, London, pp. 487-515.

Brown, S. K. and Maloney, K. E. (2009). Making sense of new apple varieties, trademarks and clubs: current status, *N. Y. Fruit Quart* 17(3), 9–12.

Brown, S. K. and Maloney, K. E. (2003). The genetic improvement of apple, *The Compact Fruit Tree* 36(2), 37–42.

Brückner, B. (2022). Measuring consumer acceptability of fruits and vegetables. In: Florkowski, W., Banks, N., Shewfelt, R. and Prussia, S. (Eds.), *Postharvest Handling: A Systems Approach* (4th edn.), Academic Press, London, pp. 651–666.

Busatto, N., Matsumoto, D., Tadiello, A., Vrhovsek, U. and Costa, F. (2019). Multifaceted analyses disclose the role of fruit size and skin-russeting in the accumulation pattern of phenolic compounds in apple, *PLoS ONE* 14(7), e0219354.

Canavari, M. (2018). Marketing research on fruit branding: the case of the pear club variety 'Angelys'. In: Cavicchi, A. and Santini, C. (Eds.) *Case Studies in the Traditional Food Sector*, Woodhead Publishing, Cambridge, pp. 239–254.

Cardello, A. V. (1995). Food quality: relativity, context and consumer expectations, *Food Qual. Prefer.* 6(3), 163–170.

Cardello, A. V., Meiselman, H. L., Schutz, H. G., Craig, C., Given, Z., Lesher, L. L. and Eicher, S. (2012). Measuring emotional responses to foods and food names using questionnaires, *Food Qual. Prefer.* 24(2), 243–250.

Chao, A. M., Loughead, J., Bakizada, Z. M., Hopkins, C. M., Geliebter, A., Gur, R. C. and Wadden, T. A. (2017). Sex/gender differences in neural correlates of food stimuli: a systematic review of functional neuroimaging studies, *Obes. Rev.* 18(6), 687–699.

Charles, M., Corollaro, M. L., Manfrini, L., Endrizzi, I., Aprea, E., Zanella, A., Grappadelli, L. C. and Gasperi, F. (2018). Application of a sensory-instrumental tool to study apple texture characteristics shaped by altitude and time of harvest, *J. Sci. Food Agr.* 98(3), 1095–1104.

Chauhan, J. and Hawrysh, Z. J. (1988). Suprathreshold sour taste intensity and pleasantness perception with age, *Physiol. Behav.* 43(5), 601–607.

Cliff, M. A. and Bejaei, M. (2018). Inter-correlation of apple firmness determinations and development of cross-validated regression models for prediction of sensory attributes from instrumental and compositional analyses, *Food Res. Int.* 106, 752–762.

Cliff, M. A., Li, J. B. and Stanich, K. (2014a). Survey reveals urban consumers' attitudes and beliefs regarding the use of wax on apples, *J. Food Qual.* 37(1), 29–41.

Cliff, M. A., Stanich, K. and Hampson, C. (2014b). Consumer research explores acceptability of a new Canadian apple – Salish™, *Can. J. Plant Sci.* 94(1), 99–108.

Cliff, M. A., Sanford, K. and Johnston, E. (1999). Evaluation of hedonic scores and R-indices for visual, flavour and texture preferences of apple cultivars by British Columbian and Nova Scotian consumers, *Can. J. Plant Sci.* 79(3), 395–399.

Cliff, M. A., Stanich, K., Lu, R. and Hampson, C. R. (2016). Use of descriptive analysis and preference mapping for early-stage assessment of new and established apples, *J. Sci. Food Agr.* 96(6), 2170–2183.

Cliff, M. A. and Toivonen, P. M. A. (2017). Sensory and quality characteristics of 'Ambrosia' apples in relation to harvest maturity for fruit stored up to eight months, *Postharvest Biol. Technol.* 132, 145–153.

Cliff, M. A., Toivonen, P. M. A., Forney, C. F., Lu, C. and Lu, C. (2010). Quality of fresh-cut apple slices stored in solid and micro-perforated film packages having contrasting O2 headspace atmospheres, *Postharvest Biol. Technol.* 58(3), 254–261.

Collinsworth, L. A., Lammert, A. M., Martinez, K. P., Leidheiser, M., Garza, J., Keener, M. and Ashman, H. (2014). Development of a novel sensory method: image Measurement of Emotion and Texture (IMET), *Food Qual. Prefer.* 38, 115–125.

Corollaro, M. L., Aprea, E., Endrizzi, I., Betta, E., Demattè, M. L., Charles, M., Bergamaschi, M., Costa, F., Biasioli, F., Grappadelli, L. C. and Gasperi, F. (2014). A combined sensory-instrumental tool for apple quality evaluation, *Postharvest Biol. Technol.* 96, 135–144.

Costa, F., Cappellin, L., Fontanari, M., Longhi, S., Guerra, W., Magnago, P., Gasperi, F. and Biasioli, F. (2012). Texture dynamics during postharvest cold storage ripening in apple (*Malus × domestica* Borkh.), *Postharvest Biol. Technol.* 69, 54–63.

Crisosto, C. H., Crisosto, G. and Neri, F. (2006). Understanding tree fruit quality based on consumer acceptance. In: Purvis, A. C., McGlasson, W. B. and Kanlayanarat, S. (Eds.), IV International Conference on Managing Quality in Chains – The Integrated View on Fruits and Vegetables Quality (vol. 712), pp. 183–190.

Cummins, S., Smith, D. M., Taylor, M., Dawson, J., Marshall, D., Sparks, L. and Anderson, A. S. (2009). Variations in fresh fruit and vegetable quality by store type, urban-rural setting and neighbourhood deprivation in Scotland, *Public Health Nutr.* 12(11), 2044–2050.

Daillant-Spinnler, B., MacFie, H. J. H., Beyts, P. K. and Hedderley, D. (1996). Relationships between perceived sensory properties and major preference directions of 12 varieties of apples from the southern hemisphere, *Food Qual. Prefer.* 7(2), 113–126.

Dalenberg, J. R., Gutjar, S., Ter Horst, G. J., de Graaf, K., Renken, R. J. and Jager, G. (2014). Evoked emotions predict food choice, *PLoS ONE* 9(12), e115388.

Deliza, R., Rosenthal, A., Hedderley, D. and Jaeger, S. R. (2010). Consumer perception of irradiated fruit: a case study using choice-based conjoint analysis, *J. Sens. Stud.* 25(2), 184–200.

Dever, M. C., Cliff, M. A. and Hall, J. W. (1995). Analysis of variation and multivariate relationships among analytical and sensory characteristics in whole apple evaluation, *J. Sci. Food Agr.* 69(3), 329–338.

Dinnella, C., Monteleone, E., Piochi, M., Spinelli, S., Prescott, J., Pierguidi, L., Gasperi, F., Laureati, M., Pagliarini, E., Predieri, S., Torri, L., Barbieri, S., Valli, E., Bianchi, P., Braghieri, A., Caro, A. D., Di Monaco, R., Favotto, S. and Moneta, E. (2018). Individual variation in prop status, fungiform papillae density, and responsiveness to taste stimuli in a large population sample, *Chem. Senses* 43(9), 697–710.

Eccher, T. and Noe, N. (1992). Influence of light on shape and quality of Golden Delicious apples. In: Lavee, S. and Goren, R. (Eds.), VII International Symposium on Plant Growth Regulators in Fruit Production (vol. 329), pp. 156–158.

Endrizzi, I., Torri, L., Corollaro, M. L., Demattè, M. L., Aprea, E., Charles, M., Biasioli, F. and Gasperi, F. (2015). A conjoint study on apple acceptability: sensory characteristics and nutritional information, *Food Qual. Prefer.* 40, 39–48.

FAOSTAT (2021). Supply utilization accounts, Available at: https://www.fao.org/faostat/en/#data/SCL; accessed February 13th, 2022.

FAOSTAT (2022). Crops and livestock products, Available at: https://www.fao.org/faostat/en/#data/QCL; accessed February 13th, 2022.

Feeney, E., O'Brien, S., Scannell, A., Markey, A. and Gibney, E. R. (2011). Genetic variation in taste perception: does it have a role in healthy eating?, *Proc. Nutr. Soc.* 70(1), 135–143.

Fenko, A., Schifferstein, H. N. and Hekkert, P. (2010). Shifts in sensory dominance between various stages of user-product interactions, *Appl. Ergon.* 41(1), 34–40.

Galmarini, M. V., Symoneaux, R., Chollet, S. and Zamora, M. C. (2013). Understanding apple consumers' expectations in terms of likes and dislikes: use of comment analysis in a cross-cultural study, *Appetite* 62, 27–36.

Gontijo, P. and Zhang, S. (2007). The mental representation of brand names: are brand names a class by themselves? In: Lowrey, T. M. (Ed.), *Psycholinguistic Phenomena in Marketing Communications* (1st edn.), Psychology Press, Baton Rogue, pp. 23–37.

Grunert, K. G., Perrea, T., Zhou, Y., Huang, G., Sørensen, B. T. and Krystallis, A. (2011). Is food-related lifestyle (FRL) able to reveal food consumption patterns in non-Western cultural environments? Its adaptation and application in urban China, *Appetite* 56(2), 357–367.

Guido, D., Perna, S., Carrai, M., Barale, R., Grassi, M. and Rondanelli, M. (2016). Multidimensional evaluation of endogenous and health factors affecting food preferences, taste and smell perception, *J. Nutr. Health Aging* 20(10), 971–981.

Gutjar, S., Dalenberg, J. R., de Graaf, C., de Wijk, R. A., Palascha, A., Renken, R. J. and Jager, G. (2015). What reported food-evoked emotions may add: a model to predict consumer food choice, *Food Qual. Prefer.* 45, 140–148.

Hampson, C. R. and McKenzie, D. L. (2006). Sensory characteristics of 21 new apple cultivars after short-term cold air storage, *J. Am. Pomol. Soc.* 60(1), 9–19.

Hampson, C. R., Quamme, H. A., Hall, J. W., MacDonald, R. A., King, M. C. and Cliff, M. A. (2000). Sensory evaluation as a selection tool in apple breeding, *Euphytica* 111(2), 79–90.

Hampson, C. R., MacDonald, R. A., McKenzie, D. L. and Walters, W. (2013). SPA493 (Salish™) apple, *Can. J. Plant Sci.* 93(5), 979-982.

Hampson, C. R., McNew, R., Cline, J., Embree, C., Embree, C., Zandstra, J. and Wilson, K. (2009). Regional differences in performance of Canadian-bred apple cultivars and implications for breeding, *Can. J. Plant Sci.* 89(1), 81–91.

Hampson, C. R., Sanford, K. and Cline, J. (2002). Preferences of Canadian consumers for apple fruit size, *Can. J. Plant Sci.* 82(1), 165–167.

Harker, F. R., Gunson, F. A. and Jaeger, S. R. (2003). The case for fruit quality: an interpretive review of consumer attitudes, and preferences for apples, *Postharvest Biol. Technol.* 28(3), 333–347.

Harker, F. R. and Hallett, I. C. (1992). Physiological changes associated with development of mealiness of apple fruit during cool storage, *Hortscience* 27(12), 1291–1294.

Harker, F. R. and Johnston, J. W. (2008). Importance of texture in fruit and its interaction with flavour. In: Brückner, B. and Wyllie, S. G. (Eds.), *Fruit and Vegetable Flavour: Recent Advances and Future Prospects*, Woodhead Publishing, Cambridge, pp.132–149.

Hein, K. A., Hamid, N., Jaeger, S. R. and Delahunty, C. M. (2012). Effects of evoked consumption contexts on hedonic ratings: a case study with two fruit beverages, *Food Qual. Prefer.* 26(1), 35–44.

Heng, Y. and House, L. A. (2018). Cluster analysis for fruit consumption patterns: an international study, *Br. Food J.* 120(9), 1942–1952.

Hoegg, J. and Alba, J. W. (2007). Linguistic framing of sensory experience: there is some accounting for taste. In: Lowrey, T. M. (Ed.), *Psycholinguistic Phenomena in Marketing Communications* (1st edn.), Routledge Taylor & Francis Group, Boca Raton, pp. 3–21.

Jacoby, J., Szybillo, G. J. and Busato-Schach, J. (1977). Information acquisition behavior in brand choice situations, *J. Consum. Res.* 3(4), 209–216.

Jaeger, S. R. (2006). Non-sensory factors in sensory science research, *Food Qual. Prefer.* 17(1–2), 132–144.

Jaeger, S. R., Andani, Z., Wakeling, I. N. and MacFie, H. J. H. (1998). Consumer preferences for fresh and aged apples: a cross-cultural comparison, *Food Qual. Prefer.* 9(5), 355–366.

Jaeger, S. R., Lusk, J. L., House, L. O., Valli, C., Moore, M., Morrow, B. and Traill, W. B. (2004). Measuring acceptance of genetically modified foods in non-hypothetical experimental markets, *Food Qual. Prefer.* 15(7–8), 401–417.

Jaeger, S. R., Machín, L., Aschemann-Witzel, J., Antúnez, L., Harker, F. R. and Ares, G. (2018). Buy, eat or discard? A case study with apples to explore fruit quality perception and food waste, *Food Qual. Prefer.* 69, 10–20.

Javornik, A. (2016). Augmented reality: research agenda for studying the impact of its medic characters on consumer behavior, *J. Retail. Consum. Serv.* 30, 252–261.

Jeong, S. and Lee, J. (2021). Effects of cultural background on consumer perception and acceptability of foods and drinks: a review of latest cross-cultural studies, *Curr. Opin. Food Sci.* 42, 248–256.

Jerald, J. (2015). *The VR Book: Human-Centered Design for Virtual Reality*, Morgan & Claypool, San Rafael, CA, pp 29–43.

Jin, S., Li, W., Cao, Y., Jones, G., Chen, J., Li, Z., Chang, Q., Yang, G. and Frewer, L. J. (2022). Identifying barriers to sustainable apple production: a stakeholder perspective, *J. Environ. Manage.* 302(B), 114082.

Jing, C., Feng, D., Zhao, Z., Wu, X. and Chen, X. (2020). Effect of environmental factors on skin pigmentation and taste in three apple cultivars, *Acta Physiol. Plant.* 42(5), 1–12.

Karagiannis, E., Michailidis, M., Tanou, G., Scossa, F., Sarrou, E., Stamatakis, G., Samiotaki, M., Martens, S., Fernie, A. R. and Molassiotis, A. (2020). Decoding altitude-activated regulatory mechanisms occurring during apple peel ripening, *Hortic. Res.* 7, 120.

Kennedy, O., Law, C., Methven, L., Mottram, D. and Gosney, M. (2010). Investigating age-related changes in taste and affects on sensory perceptions of oral nutritional supplements, *Age Ageing* 39(6), 733–738.

Keller, K. L. (2003). Understanding brands, branding and brand equity, *J. Interact. Mark.* 5(1), 7–20.

Keller, K. L. and Swaminathan, V. (2020). *Strategic Brand Management: Building, Measuring, and Managing Brand Equity* (5th edn.), Pearson, Harlow.

Kenney, E. and Adhikari, K. (2016). Recent developments in identifying and quantifying emotions during food consumption, *J. Sci. Food Agr.* 96(11), 3627–3630.

Khan, M. A. and Hackler, L. R. (1981). Evaluation of food selection patterns and preferences, *Crit. Rev. Food Sci. Nutr.* 15(2), 129–153.

Khanizadeh, S., Groleau, Y., Levasseur, A., Charles, M. T., Tsao, R., Yang, R., DeEll, J., Hampson, C. R. and Toivonen, P. (2007). Eden(TM) a non-browning apple suitable for fresh cut slice and fruit salad. In: XII EUCARPIA Symposium on Fruit Breeding and Genetics (vol. 814), pp. 273–274.

Kingston, C. M. (1992). Maturity indices for apple and pear. In: Janick, J. (Ed.), *Horticultural Reviews* (vol. 13), John Wiley & Sons, New York, pp. 407–432.

Kong, Y., Sharma, C., Kanala, M., Thakur, M., Li, L., Xu, D., Harrison, R. and Torrico, D. D. (2020). Virtual reality and immersive environments on sensory perception of chocolate products: a preliminary study, *Foods* 9(4), 515.

Köster, E. P. (2009). Diversity in the determinants of food choice: a psychological perspective, *Food Qual. Prefer.* 20(2), 70–82.

Koutsimanis, G., Getter, K., Behe, B., Harte, J. and Almenar, E. (2012). Influences of packaging attributes on consumer purchase decisions for fresh produce, *Appetite* 59(2), 270–280.

Lancaster, J. E. and Dougall, D. K. (1992). Regulation of skin color in apples, *Crit. Rev. Plant Sci.* 10(6), 487–502.

Lancaster, J. E., Grant, J. E., Lister, C. E. and Taylor, M. C. (1994). Skin color in apples–influence of copigmentation and plastid pigments on shade and darkness of red color in five genotypes, *J. Am. Soc. Hortic. Sci.* 119(1), 63–69.

Lange, C., Issanchou, S. and Combris, P. (2000). Expected versus experienced quality: trade-off with price, *Food Qual. Prefer.* 11(4), 289–297.

Lefebvre, A., Bassereau, J. F., Pensé-Lheritier, A. M., Rivère, C., Harris, N. and Duchamp, R. (2010). Recruitment and training of a sensory expert panel to measure the touch of beverage packages: issue and methods employed, *Food Qual. Prefer.* 21(1), 156–164.

Lichtenstein, D. R., Ridgway, N. M. and Netemeyer, R. G. (1993). Price perceptions and consumer shopping behavior: a field study, *J. Mark. Res.* 30(2), 234–245.

Lombart, C., Millan, E., Normand, J. M., Verhulst, A., Labbé-Pinlon, B. and Moreau, G. (2019). Consumer perceptions and purchase behavior toward imperfect fruits and vegetables in an immersive virtual reality grocery store, *J. Retail. Consum. Serv.* 48, 28–40.

Lund, C. M., Jaeger, S. R., Amos, R. L., Brookfield, P. and Harker, F. R. (2006). Tradeoffs between emotional and sensory perceptions of freshness influence the price consumers will pay for apples: results from an experimental market, *Postharvest Biol. Technol.* 41(2), 172–180.

Lusk, J. L. (2003). Using experimental auctions for marketing applications: a discussion, *J. Agric. Appl. Econ.* 35(2), 349–360.

Lüscher, P., Laube, P., Hägi, S., Rogers, S. and Flitter, H. (2013). Weighting by pairwise comparison, *Geographic Information Technology Training Alliance*. Available at: http://www.gitta.info/Suitability/en/html/Normalisatio_learningObject3.html; accessed February 13th, 2022.

Ma, G. (2015). Food, eating behavior, and culture in Chinese society, *J. Ethn. Foods* 2(4), 195–199.

McCabe, D. B. and Nowlis, S. M. (2003). The effect of examining actual products or product descriptions on consumer preference, *J. Consum. Psychol.* 13(4), 431–439.

McMahon, E. J., Campbell, K. L. and Bauer, J. D. (2014). Taste perception in kidney disease and relationship to dietary sodium intake, *Appetite* 83, 236–241.

Meißner, M., Pfeifferb, J., Peukertc, C., Dietrichd, H. and Pfeiffere, T. (2020). How virtual reality affects consumer choice, *J. Bus. Res.* 117, 219–231.

Published by Burleigh Dodds Science Publishing Limited, 2024.

Meiselman, H. L. (2007). Integrating consumer responses to food products. In: MacFie, H. J. H. (Ed.), *Consumer Led Food Product Development*, Woodhead Publishing, Cambridge, pp. 3-33.

Mennella, J. A., Jagnow, C. P. and Beauchamp, G. K. (2001). Prenatal and postnatal flavor learning by human infants, *Pediatrics* 107(6), E88.

Mojet, J., Christ-Hazelhof, E. and Heidema, J. (2001). Taste perception with age: generic or specific losses in threshold sensitivity to the five basic tastes?, *Chem. Senses* 26(7), 845-860.

Mojet, J., Christ-Hazelhof, E. and Heidema, J. (2005).Taste perception with age: pleasantness and its relationships with threshold sensitivity and supra-threshold intensity of five taste qualities, *Food Qual. Prefer.* 16(5), 413-423.

Mojet, J., Heidema, J. and Christ-Hazelhof, E. (2003). Taste perception with age: generic or specific losses in supra-threshold intensities of five taste qualities?, *Chem. Senses* 28(5), 397-413.

Montero-Vicente, L., Roig-Merino, B., Buitrago-Vera, J. and Sigalat-Signes, E. (2019). Characterisation of fresh fruit consumption in Spain based on food-related lifestyle, *Br. Food J.* 121(12), 3307-3320.

Murphy, C. (1993). Nutrition and chemosensory perception in the elderly, *Crit. Rev. Food Sci. Nutr.* 33(1), 3-15.

Musacchi, S. and Serra, S. (2018). Apple fruit quality: overview on pre-harvest factors, *Sci. Hortic.* 234, 409-430.

Nestle, M., Wing, R., Birch, L., Disogra, L., Drewnowski, A., Middleton, S., Sigman-Grant, M., Sobal, J., Winston, M. and Economos, C. (2009). Behavioral and social influences on food choice, *Nutr. Rev.* 56(5), 50-64.

Ng, M., Chaya, C. and Hort, J. (2013). The influence of sensory and packaging cues on both liking and emotional, abstract and functional conceptualisations, *Food Qual. Prefer.* 29(2), 146-156.

Nicklaus, S., Boggio, V., Chabanet, C. and Issanchou, S. (2005). A prospective study of food variety seeking in childhood, adolescence and early adult life, *Appetite* 44(3), 289-297.

Olegario, L. S., Estevéz, M., González-Mohino, A., Madruga, M. S. and Ventanas, S. (2021). Cross-cultural emotional response to food stimuli: influence of consumption context, *Food Res. Int.* 142, 110194.

Olson, J. C. and Jacoby, J. (1972). Cue utilization in the quality perception process. In: Venkatesan, M. (Ed.), *SV - Proceedings of the Third Annual Conference of the Association for Consumer Research*, Association for Consumer Research, Chicago, IL, pp. 167-179.

Ossadnik, W., Schinke, S. and Kaspar, R. H. (2016). Group aggregation techniques for analytic hierarchy process and analytic network process: a comparative analysis, *Group Decis. Negot.* 25(2), 421-457.

Pangborn, R. M. and Pecore, S. D. (1982). Taste perception of sodium chloride in relation to dietary intake of salt, *Am. J. Clin. Nutr.* 35(3), 510-520.

Pilone, V., Stasi, A. and Baselice, A. (2017). Quality preferences and pricing of fresh-cut salads in Italy: new evidence from market data, *Br. Food J.* 119(7), 1473-1486.

Piqueras-Fiszman, B. and Jaeger, S. R. (2015). The effect of product-context appropriateness on emotion associations in evoked eating occasions, *Food Qual. Prefer.* 40(A), 49-60.

Piqueras-Fiszman, B. and Spence, C. (2012). The influence of the feel of product packaging on the perception of the oral-somatosensory texture of food, *Food Qual. Prefer.* 26(1), 67–73.

Pramudya, R. C. and Seo, H. S. (2019). Hand-feel touch cues and their influences on consumer perception and behavior with respect to food products: a review, *Foods* 8(7), 259.

Proserpio, C., Laureati, M., Invitti, C., Cattaneo, C. and Pagliarini, E. (2017). BMI and gender related differences in cross-modal interaction and liking of sensory stimuli, *Food Qual. Prefer.* 56, 49–54.

Ranatunga, C. L., Jayaweera, H. H. and Ariyaratne, T. R. (2008). Evaluation of finger-feel firmness as a subjective measurement of tomato quality degradation in the retail market, *Trop. Agr. Res.* 20, 134–142.

Raudenbush, B., Schroth, F., Reilley, S. and Frank, R. A. (1998). Food neophobia, odor evaluation and exploratory sniffing behavior, *Appetite* 31(2), 171–183.

Reganold, J. P., Glover, J. D., Andrews, P. K. and Hinman, H. R. (2001). Sustainability of three apple production systems, *Nature* 410(6831), 926–930.

Richards, T. J. and Patterson, P. M. (2000). New varieties and their returns to commodity promotion: the case of Fuji apples, *Agric. Resour. Econ. Rev.* 29(1), 10–23.

Rickard, B. J., Schmit, T. M., Gomez, M. I. and Lu, H. (2013). Developing brand names for patented fruit varieties: does the name matter?, *Agribusiness* 29(3), 259–272.

Romeo-Arroyo, E., Mora, M., Pazos, N., Deba-Rementeria, S. and Vázquez-Araújo, L. (2022). Effect of product properties and context on the perception of sweetness and liking: a case study with butter cookies, *J. Sens. Stud.* 37(3), e12740.

Rosa, A., Pinna, I. and Masala, C. (2022). Role of body weight and sex in the olfactory and gustatory pleasantness, intensity, and familiarity of a lipid-rich food, *J. Sens. Stud.* 37(3), e12739.

Samant, S. S., Chapko, M. J. and Seo, H. S. (2017). Predicting consumer liking and preference based on emotional responses and sensory perception: a study with basic taste solutions, *Food Res. Int.* 100(1), 325–334.

Sánchez-Bravo, P., Chambers, E., Noguera-Artiaga, L., López-Lluch, D., Carbonell-Barrachina, Á. A. and Sendra, E. (2020). Consumers attitude towards the sustainability of different food categories, *Foods* 9(11), 1608.

Schürmann, M., Caetano, G., Hlushchuk, Y., Jousmäki, V. and Hari, R. (2006). Touch activates human auditory cortex, *Neuroimage* 30(4), 1325–1331.

Slocombe, B. G., Carmichael, D. A. and Simner, J. (2016). Cross-modal tactile-taste interactions in food evaluations, *Neuropsychologia* 88, 58–64.

Statista (2021). Ready-to-eat meals, Available at: https://www.statista.com/outlook/cmo/food/convenience-food/ready-to-eat-meals/worldwide; accessed February 13th, 2022.

Strohm, K. (2013). Of the 30,000 apple varieties found all over the world only 30 are used and traded commercially, *Agribenchmark*, Available at: http://www.agribenchmark.org/agri-benchmark/did-you-know/einzelansicht/artikel//only-5500-wi.html; accessed February 13th, 2022.

Sulmont-Rossé, C., Drabek, R., Almli, V. L., van Zyl, H., Silva, A. P., Kern, M., McEwan, J. A. and Ares, G. (2019). A cross-cultural perspective on feeling good in the context of foods and beverages, *Food Res. Int.* 115, 292–301.

Summerland Varieties Corp. (2022). Our varieties, Available at: https://www.summerlandvarieties.com/varieties/; accessed March 18th, 2022.

Szczesniak, A. S. and Bourne, M. C. (1969). Sensory evaluation of food firmness, *J. Texture Stud.* 1(1), 52-64.

Tamers, S. L., Agurs-Collins, T., Dodd, K. W. and Nebeling, L. (2009). US and France adult fruit and vegetable consumption patterns: an international comparison, *Eur. J. Clin. Nutr.* 63(1), 11-17.

Teh, S. L., Brutcher, L., Schonberg, B. and Evans, K. (2020). Eleven-year correlation of physical fruit texture traits between computerized penetrometers and sensory assessment in an apple breeding program, *HortTechnology* 30(6), 719-724.

Teh, S. L., Kostick, S., Brutcher, L., Schonberg, B., Barritt, B. and Evans, K. (2021). Trends in fruit quality improvement from 15 years of selection in the apple breeding program of Washington State University, *Front. Plant Sci.* 12, 714325.

Tijskens, P. and Schouten, R. (2022). Modeling quality attributes and quality-related product properties. In: Florkowski, W., Banks, N., Shewfelt, R. and Prussia, S. (Eds.), *Postharvest Handling: A Systems Approach* (4th edn.), Academic Press, London, pp. 99-133.

Tuorila, H., Meiselman, H. L., Bell, R., Cardello, A. V. and Johnson, W. (1994). Role of sensory and cognitive information in the enhancement of certainty and liking for novel and familiar foods, *Appetite* 23(3), 231-246.

Vakratsas, D. and Ambler, T. (1999). How advertising works: what do we really know?, *J. Mark.* 63(1), 26-43.

Van Herpen, E., van den Broek, E., van Trijp, H. C. M. and Yu, T. (2016). Can a virtual supermarket bring realism into the lab? Comparing shopping behavior using virtual and pictorial store representations to behavior in a physical store, *Appetite* 107, 196-207.

Ventura, A. K. and Worobey, J. (2013). Early influences on the development of food preferences, *Curr. Biol.* 23(9), R401-R408.

Wang, W., Celton, J. M., Buck-Sorlin, G., Balzergue, S., Bucher, E. and Laurens, F. (2020). Skin color in apple fruit (*Malus* × *domestica*): genetic and epigenetic insights, *Epigenomes* 4(3), 13.

Wang, Q. J., Barbosa Escobar, F., Alves Da Mota, P. and Velasco, C. (2021). Getting started with virtual reality for sensory and consumer science: current practices and future perspectives, *Food Res. Int.* 145, 110410.

Wang, Q. J., Mielby, L. A., Junge, J. Y., Bertelsen, A. S., Kidmose, U., Spence, C. and Byrne, D. V. (2019). The role of intrinsic and extrinsic sensory factors in sweetness perception of food and beverages: a review, *Foods* 8(6), 211.

World Bank. (2021). Fruit, edible; apples, fresh exports by country in 2019, *World Integrated Trade Solutions*, Available at: https://wits.worldbank.org/trade/comtrade/en/country/ALL/year/2019/tradeflow/Exports/partner/WLD/product/080810#; accessed February 13th, 2022.

Zaichkowsky, J. L. and Vipat, P. (1993). Inferences from brand names. In: Van Raaij, W. F. and Bamossy, G. J. (Eds.), *European Advances in Consumer Research* (vol. 1), Association for Consumer Research, Provo, UT, pp. 534-540.

Zeithaml, V. A. (1988). Consumer perceptions of price, quality, and value: a means-end model and synthesis of evidence, *J. Mark.* 52(July), 2-22.

Chapter 2

Advances in understanding texture development in apples

Hilde Nybom, Swedish University of Agricultural Sciences, Sweden

1 Introduction

While apple cultivars were historically appreciated mainly for their taste attributes, such as sweetness, balanced acidity and complex aroma, modern-day consumers also rate fruit texture as important. Texture is, however, a complex concept, and a number of interrelated traits have been identified in consumer tests, including chewiness, cohesiveness, crispness, crunchiness, fibrousness, firmness, hardness, juiciness, mushiness and mealiness (Charles et al., 2018). Sensory panel-based analyses have shown that high values are generally desired for firmness, crispness and juiciness, while mealiness should be as low as possible (Daillant-Spinnler et al., 1996; Vigneau et al., 2014).

Hardness and firmness are often used as synonyms and can be interpreted as the force exerted when biting through a piece of fruit, thereby causing breakage of cell walls as well as affecting cell turgor. By contrast, crispness is often associated with the amount of acoustic energy released when the cell walls are broken. The intracellular juice released by the broken cell walls give

http://dx.doi.org/10.19103/AS.2023.0127.03

rise to juiciness. Biting into a mealy (floury, starchy) fruit with soft, dry and/or granular fruit flesh causes cells along the middle lamellae of the cell walls to slip and slide without breaking the cell walls. This dampens the sound and lowers the amount of juice released.

Since a large proportion of marketed fruit is stored for several months before it reaches consumers, good texture is important not just at harvest but also following cold storage and during the 1 to 2 weeks at ambient temperatures before the fruit is consumed. Loss of firmness during storage is caused by depolymerization (degradation) of the cell wall middle lamellae polysaccharide structure (Brummell and Harpster, 2001). A dry and mealy texture is related to strong depolymerization of the cell walls, while cultivars with firm and crispy fruit retain cell wall integrity to a greater extent (Longhi et al., 2013).

Storage capability (storability) is extremely important to the grower's economy and must be taken into consideration when deciding what cultivars to grow, and what techniques to apply to growing, harvesting and storing the fruit. Firm fruits tend to tolerate mechanical damage during harvesting and transportation and endure long-term storage as well as attacks from fungal diseases.

There are concerns that some original apple flavors may be lost in newer superfirm cultivars. This chapter will address recent findings related to the concept of fruit texture and how to improve this in new apple cultivars.

2 Anatomy and physiology of fruit texture traits

Fruit anatomy and chemical processes have a significant influence on the apple's textural properties. Although samples of a single apple cultivar ('Golden Delicious') did not show a correlation between size and number of cells in the fruit flesh and variation in fruit flesh texture (Charles et al., 2018), other studies with several different cultivars have indicated a strong relationship between variation in fruit texture and the size and shape of the cells. Smaller intercellular spaces usually co-occur with rounded cells, while fruit flesh with angular cells has larger intercellular spaces. Firm and hard apples tend to have rounded cells and smaller intercellular spaces than softer apples, resulting in more densely packed tissue (Li et al., 2019). Poles et al. (2020) showed that cell size was a more important predictor than cell shape and that smaller cells result in firmer fruit.

Mann et al. (2005) suggested that fruits with a lower number of cells per unit area are crispier than fruits with a higher number and that cell size could be a good predictor of juiciness since bigger cells release more juice than smaller ones, but this has been refuted in other studies. Cell size appeared to have little impact on juiciness, with cell shape being more important; cultivars with juicy fruit tend to have rounded cells instead of more angular cells (Poles et al., 2020).

Since apples are a climacteric fruit, loss of firmness (fruit softening) is related to the internal level of ethylene, which regulates cell wall degradation (Wakasa et al., 2006; Wei et al., 2010). Fruit ripening-related softening starts during development from fruitlet to the fully ripened stage and is characterized by a pronounced degradation of the pectin in cell walls. This process starts with the loss of the side-chain neutral sugars galactose and arabinose through action of β-galactosidase and α-L-arabinofuranosidase, respectively (Gwanpua et al., 2018). Pectin polysaccharides with a low degree of branching are then demethoxylated by pectin methylesterase. Finally, the demethoxylated pectin chains are attacked by polygalacturonase, resulting in depolymerization of the cell wall pectin.

The softer fruit typical of 'Royal Gala' has been shown to lose more galactose compared to the firmer fruit of 'SciFesh', which has lower β-galactosidase activity and higher cell wall galactan content and thus may better withstand cell wall-modifying enzymes (Ng et al., 2015). Activity of β-galactosidase is higher and pectin degradation is more prominent in 'Fuji', which has crisp fruits that soften markedly during storage, compared to the firm and tough fruits of 'Qinguan', which retain initial firmness during the entire storage period (Yang et al., 2018). Another study showed that galactose and arabinose contribute to the higher hardness of 'Hanfu', while arabinose, egg-box structure and fucosylated xyloglucans improve cell adhesion and thus contribute to the higher crispness of 'Honeycrisp' (Yang et al., 2022).

Generally, firmness decreases during the harvesting period, with a major reduction in conjunction with or, more commonly, just after the climacteric rise in IEC (internal ethylene concentration; Tahir and Nybom 2013). Commercial fruit, especially if intended for long-term storage, is usually harvested just before the rise in IEC. Harvesting the fruit later leads to faster deterioration of overall quality. In an experiment, when 'Golden Delicious' was harvested 1 (T1) or 3 weeks (T2) after optimal harvesting time (T0), the T2 samples especially were less hard and crunchy and more mealy and grainy compared to T0 and T1 samples, according to both instrumental and sensory panel data (Charles et al., 2018).

Even when harvested at the optimal stage, fruit texture changes conspicuously during storage. For germplasm evaluation and selection of breeding materials, fruit texture parameters are therefore often measured both at harvest and after storage. Loss in firmness (difference in firmness between measurements) is sometimes divided by number of weeks in storage to yield 'softening rate' (Nybom et al., 2013). In addition, measurements are sometimes conducted following removal of the fruit from storage in order to assess the shelf-life, i.e. the period when the fruit is exposed to ambient temperatures before purchase and consumption.

The ability to maintain quality throughout a period of cold storage is most likely influenced by differences in ethylene production as well as cell size and cell wall structures. Early-ripening cultivars (summer and early autumn apples) generally have high climacteric respiration and higher levels of ethylene production and therefore mature and soften quickly (Janisiewicz et al., 2008). By contrast, late-ripening cultivars (late autumn and winter apples) have lower respiration and ethylene production rates, they mature more slowly and retain firmness better during storage. In addition, early-ripening cultivars generally have larger cells and larger intercellular spaces compared to late-ripening ones (Johnston et al., 2002).

3 Evaluation of fruit texture parameters

Sensory evaluations of various different textural traits have been carried out on apples, using both mass-testing on hundreds of untrained consumers (Jönsson and Nybom 2007) as well as trained panels comprising 10–30 respondents (Daillant-Spinnler et al., 1996; Hampson et al., 2000). To untrained consumers, texture can be hard to evaluate since the difference between firmness and crispness is not always fully understood. Evaluations with trained sensory panelists have, however, also yielded variable results (Teh et al., 2020). Sensory crispness is generally moderately correlated with sensory juiciness, and sometimes with hardness, but the set of apple cultivars tested differs widely between studies, as do the training and terminology used by panelists. Moreover, the number of samples that can be assessed by the same taste panel within a restricted time frame is quite small.

Instrumentally derived data are usually less expensive and can be obtained in very large numbers, thereby providing the sample sizes needed for in-depth physiological and genetic studies. Relationships between sensory panel variables and instrumental measurements must, however, be carefully defined (Bejaei et al., 2021). Predictive models using both sensory and instrumental data were shown to explain more than 85% of the variation for hardness and crispness, but accurate models for juiciness and skin toughness were more difficult to achieve (Fig. 1).

Firmness (i.e. hardness) is the most commonly investigated parameter and is easily assessed in the field by a puncture or pressure test administered by a simple handheld penetrometer (e.g. Effegi) using a metal probe, 8–11 mm in diameter with a mildly convex or flat tip. The difference between peeled (most common) and unpeeled samples must be taken into consideration since the skin can contribute around 60% of overall firmness (Grotte et al., 2001; Costa, 2016). For large-scale research data, the probe is often attached to an electronic device that allows linear, constant movement of the crosshead with the probe, e.g. Instron. Frequently, these measurements over-emphasize the

Published by Burleigh Dodds Science Publishing Limited, 2024.

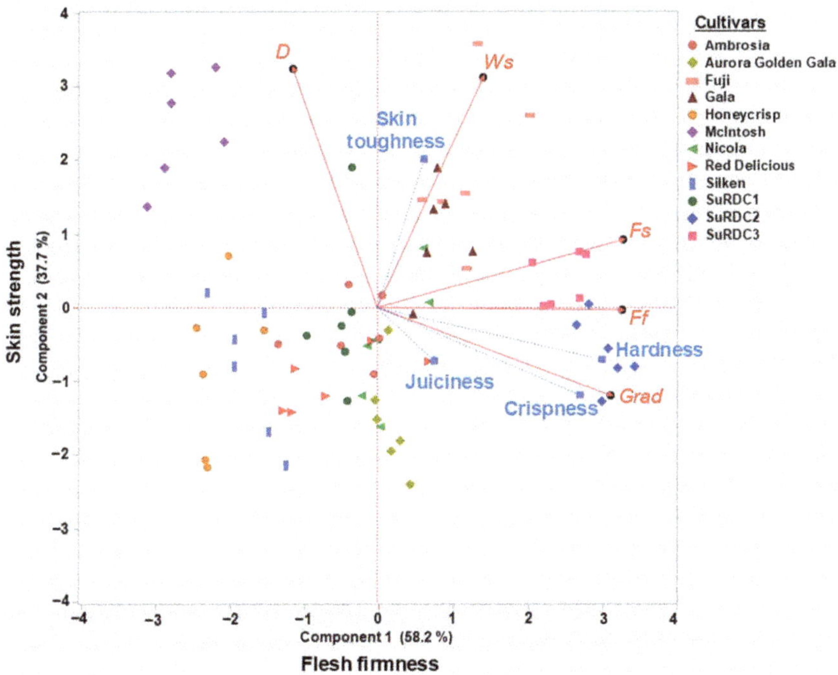

Figure 1 Principal component analysis biplot calculated using standardized mean values of 5 TA.XT*plus* Texture Analyzer parameters (*Fs*, *Ws*, *Grad*, *D* and *Ff*). Instrumental parameters are identified by red lines. Samples of the 12 apple cultivars/selections are identified with symbols of unique colors and shapes. Sensory attributes (crispness, hardness, juiciness and skin toughness) are identified by blue lines and positioned on this plot using correlation analysis. Source: Reproduced with permission from Bejaei et al. (2021).

influence of the outer cortex of the fruit since the probe does not reach the inner parts.

Since crispness is related to the sound emitted when eating an apple, various acoustic methods have been applied to assess this parameter. The instrumental acoustic-impulse response technique consists of administrating a gentle tap on the fruit and measuring the frequency of the sound within the audible spectrum. The obtained readings are strongly associated with the water content and turgor of the fruit. An automated texture analyzer (e.g. TA.XT Texture Analyzer) equipped with an acoustic envelope device measures both mechanical and acoustic parameters on the same sample. Data from this type of instrument has proven to yield superior results compared to measuring the acoustic response from human testers actually biting into the fruit (Piazza and Giovenzana, 2015).

When applied to fruit at harvest, penetrometer-derived estimates of firmness may show little association with acoustic-based estimates of crispness (Gwanpua et al., 2015; Sadar et al., 2018). However, both methods produce progressively lower values when applied to stored fruit, especially during the first months of storage (Sadar et al., 2018). When applied to shelf-life assessments, acoustics-based values indicated a consistent decrease in crispness as expected, while penetrometer values were more unpredictable (Sadar et al., 2018). The attachment of carboxyl groups arising from pectin degradation can lead to a drier and more rubbery fruit texture (Gwanpua et al., 2015), sometimes producing constant or even higher penetrometer estimates of peel hardness and fruit flesh firmness in stored fruit compared to fruit at harvest (Nybom et al., 2013; Spoor et al., 2019; Butkeviciute et al., 2021).

An informative overview of studies on correlations between instrumental data and sensory panel-based evaluations of various apple texture parameters has been provided by Kim et al. (2022). A strong positive association has usually been found with perceived firmness, as well as with crispness, crunchiness, hardness and juiciness, and a weak negative association with mealiness. Generally, a combination of mechanical and acoustic parameters show significant associations with sensory panel-derived scores for hardness, crispness, juiciness and mealiness when different apple cultivars are investigated (Zdunek et al., 2011; Corollaro et al., 2014; Ting et al., 2015; Charles et al., 2018). When analyzing 86 apple cultivars with a texture analyzer, mechanical impact-based parameters showed high correlation to firmness, while acoustics-based parameters corresponded to crispness as perceived by human senses (Costa et al., 2011, 2012). High crispness always co-occurs with high firmness, whereas apples with low crispness can have any level of firmness (Costa et al., 2011; Ting et al., 2015). In another study, it was shown that the same cultivar may have high values for parameters associated mainly with firmness but low values for parameters associated mainly with crispness (Poles et al., 2020).

An especially large study, based on 11 years of routine fruit quality evaluations with both instrumental and sensory panel data, was carried out in an American apple breeding program (Teh et al., 2020). A Mohr Digi-Test computerized penetrometer, which permits the testing of different layers of fruit, was used to assess hardness (force encountered by the test plunger) and crispness (energy released during the fruit tearing). Instrumental hardness traits significantly correlated with sensory perceptions of hardness, while a lower but still significant correlation was found between instrumental crispness values and the sensory evaluation of crispness, which was influenced also by juiciness. In another study, three different penetrometers (Fruit Texture Analyzer, Mohr Digi-Test-2 and TA.XTplus Texture Analyzer) were applied to the same eight cultivars and associations between the different sets of instrumental data and a set of sensory data were investigated, and models developed for converting data between the different instruments (Bejaei, 2022).

Attaching a friction rig with polypropylene balls to a texture analyzer helped yield data that correlated with various fruit texture parameters (Kim et al., 2022). Friction coefficients were positively correlated with sensory panel-derived scores for crispness and juiciness, and negatively correlated with scores for mealiness and 'rate of melt'. Optical measurements have also been used to assess firmness; time-resolved reflectance spectroscopy in the 580–1064 nm wavelength region showed the potential for discrimination between cultivars (Vanoli et al., 2018a). Yet another option is aquaphotomics based on investigating water-light interactions with NIR (Near Infra-Red) spectrometry (Vanoli et al., 2018b). Hydrolyzation of the pectin apparently affects water structures in the fruit and thus produces water spectral patterns that change according to differences in fruit texture.

4 Influence of growing conditions

Apples are grown in a wide range of climates and experience major differences in terms of quality and quantity of light, temperature, moisture and day length. Fruit from Northern lowland areas (Belgium) had a firmness corresponding to fruit grown at 1000 m in South Tyrol (Sadar et al., 2018), indicating that cultivars developed for colder areas may become too soft when grown in a warmer climate. 'Braeburn' and 'Kanzi' varieties harvested at two different altitudes (300 m and 650 m) and stored in commercial controlled atmosphere (CA) storage for 9 months were shown to differ substantially in firmness (Tijskens et al., 2018). It is likely that lower growing temperatures at higher altitudes reduce the cell division rate during the early growth phase, which results in smaller and firmer apples. Fruit from the same cultivar but grown at higher altitudes similarly had higher initial firmness compared to fruit grown at lower altitudes (Sadar et al., 2018). In another set of analyses, fruit from lower altitudes were juicier, crunchier and sweeter compared to high-altitude samples, which were described as more mealy, sour and astringent by a sensory panel (Charles et al., 2018). Texture performance, soluble solids content and titratable acidity corroborated this sensory description. Moreover, anatomical data showed that fruit from lower altitudes had a larger volume, a higher number of cells and a higher percentage of intercellular spaces.

Commonly applied production systems, i.e. organic, conventional and integrated, provide different environmental conditions for fruit trees and may consequently affect fruit texture, even on a local scale. In a set of studies in Washington state in the United States, organically grown apples were shown to be significantly firmer than same-sized fruit in other production systems (Reganold et al., 2001; Peck et al., 2006). Even small changes in irrigation, mulching, pruning and fertilization have been shown to affect fruit firmness at harvest and/or after storage and/or after 1 to 2 weeks of shelf-life testing

(Mpelasoka et al., 2001; Tahir and Gustavsson, 2010; Tahir et al., 2015b). Differences in crop load can also affect firmness, likely due to variations in the amount of available photoassimilates per fruit (Serra et al., 2016; Tijskens et al., 2018).

5 Fruit texture and storability

The effects of different storage conditions on fruit quality, including nutritional value, taste and texture (especially firmness) have been studied in many apple cultivars. For early-ripening cultivars, a desirable texture, i.e. crispy and medium firm, is needed at harvest since the fruit is consumed within a few weeks. Medium- and late-ripening cultivars need to retain a desirable texture for much longer since the fruit is stored for several months and even up to a year in modern facilities with CA.

Despite cultivar-dependent optimization of storage conditions, firmness, crispness and juiciness usually decrease during cold storage. Treatment of stored apples with 1-methylcyclopropene (1-MCP) is an efficient way to delay fruit softening and is now used in many parts of the world, although not for organically grown fruit. Recently, 1-MCP treated fruit of 'Hwangok' and 'Picnic' varieties, stored up to 6 months at 0°C, were shown to maintain firmness and exhibit lower internal ethylene concentrations compared with untreated fruit (Win et al., 2021). Analyses of both treated and untreated samples suggest that 1-MCP maintained cell wall pectin and delayed softening by reducing solubilization of polyuronides and neutral sugars and limiting cell wall hydrolysis due to ethylene-dependent processes.

Some of the inherent differences in storability can be explained by inter-cultivar differences in texture dynamics (Costa et al., 2012). Three major trends have been identified in 83 cultivars following a comparison of texture analyzer data taken both at harvest and after 2 months of cold storage. Cultivars such as 'Golden Delicious' and other older varieties showed a general decrease in both mechanical and acoustic profiles during storage. By contrast, cultivars like 'Maigold' remained stable throughout the entire storage period. Finally, a set of cultivars including 'Fuji' showed a slight increase, especially in acoustic parameters. In general, apples lose density and increase in volume during storage, resulting in higher intercellular air fraction. A high acoustic response following storage can perhaps be explained by a combination of a high air fraction, high turgor and high integrity of the middle cell wall lamellae.

Ripening period (number of days after flowering to optimum harvest date) was positively correlated with firmness at harvest and negatively correlated with softening rate in 127 Swedish-grown cultivars (Nybom et al., 2013). Corresponding associations between ripening period, firmness at harvest and softening rate have also been reported in other studies (Tahir et al., 2015a;

Migicovsky et al., 2021). In addition, polyploid cultivars have displayed significantly greater firmness at harvest compared to diploid cultivars but similar softening rates (Nybom et al., 2013). Interestingly, fruits with a lower number of seeds (typical also for polyploid apples) have higher firmness than fruit with many seeds, possibly due to a variation in ethylene production (Buccheri and Di Vaio, 2005).

6 Fruit texture and fungal diseases

Variation in storability is closely associated with resistance to fungal storage diseases (Blažek et al., 2007). This factor becomes especially important in integrated and organic production where chemical fungicides are restricted or even prohibited. The resulting lack of protection enhances susceptibility to these fungi, and a substantial amount of fruit is destroyed each year. Most of this damage is caused by ascomycete fungi, known as storage rots, which attack apples both in the orchard and during cold storage (see review in Nybom et al., 2020). Symptoms are first visible as lesions on the fruit epidermis and can proceed to rotting of the entire fruit. Research data on susceptibility to different rots is based mainly on experimental inoculations, usually involving the transfer of conidiospores into apple flesh with a micropipette (wound-inoculation). The diameter of the resulting lesion is measured after several weeks of cold storage and used as a measure of susceptibility.

Some of the most significant storage rots in apples belong to two biotrophic or hemibiotrophic genera also known as latent infection pathogens, namely *Neofabraea* (= *Pezicula*) and *Colletotrichum*. Apple cultivars assessed from natural infection to be resistant to *Neofabraea* had somewhat firmer flesh on average than susceptible cultivars (Blažek et al., 2007) but no correlation was found with firmness when dipping fruit of 18 cultivars in a spore suspension of *Colletotrichum acutatum* (Biggs and Miller, 2001). A correlation between the amount of softening during storage and susceptibility was found after wound-inoculating with *Colletotrichum gloeosporioides* in a set of 36 early-ripening cultivars, although not for 34 late-ripening ones (Ahmadi-Afzadi et al., 2013). Neither set of cultivars showed any impact on firmness at harvest. Although fruit texture may play a role, tolerance to these fungi depends mainly on the number of lenticels (major entry points) and the thickness of the cuticular layer of the fruit, as well as various resistance genes, although none has as yet been identified.

Some mainly necrotrophic fungal species are known as 'wound pathogens' and include, among others, *Penicillium expansum*, *Botrytis cinerea*, *Monilinia fructigena* and *Monilinia laxa*. One QTL (quantitative trait loci) for relatively strong resistance toward the most well-researched of these fungi, namely *P. expansum* which causes blue mold, has been identified on linkage group (LG) 3 in the apple's

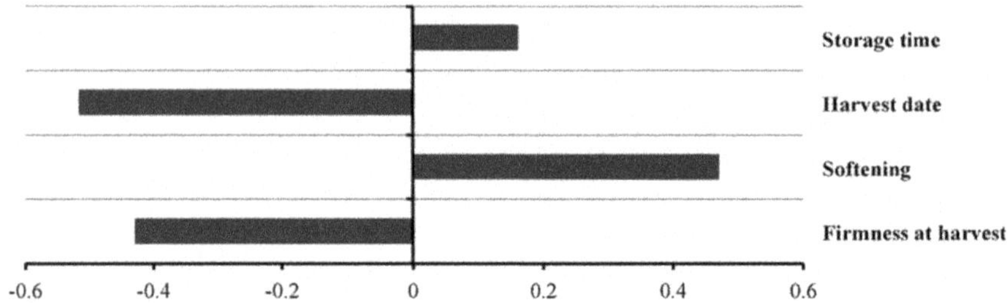

Figure 2 Weighted regression coefficient indicating the importance of four independent variables: storage time (number of weeks in storage), harvest date (ripening time), softening and firmness at harvest in explaining variation in size of lesions produced after artificial inoculation of 92 Swedish-grown apple cultivars with *P. Expansum*. Source: Reproduced with permission from Ahmadi-Afzadi et al. (2013).

wild relative *Malus sieversii*, while another QTL on LG 10, co-occurring with QTLs for firmness and ripening, provided lower-level resistance and was presumably inherited from 'Royal Gala' (Norelli et al., 2017). More research is required to unravel the genetic structure among these QTLs and their interdependence.

Animal- or man-made wounds in the fruit constitute major entry points for wound pathogens. Several experiments point to the role of fruit flesh texture in terms of tolerance to these fungi, reflecting the ease with which an infection point can lead to a large lesion. Regression analyses with *P. expansum* lesion diameter as a dependent variable demonstrated a negative effect of fruit firmness at harvest in 46 late-ripening cultivars but not in 46 early-ripening ones (Ahmadi-Afzadi et al., 2013). The amount of fruit softening during storage had a positive effect on lesion diameter in late-ripening cultivars but not in early-ripening (Fig. 2). In a follow-up study on 81 apple cultivars, lesion diameter was again negatively associated with fruit firmness at harvest and positively associated with the amount of fruit softening during storage (Tahir et al., 2015a). Similar results were reported in a smaller study by Costa et al. (2005). A negative correlation was also found between fruit firmness and lesion diameter in a study of Iranian cultivars (Naeem-Abadi et al., 2014). The relationship between firmness and lesion decay reported here is likely associated with the ability of cell walls to withstand attacks from pectolytic enzymes of the fungus.

Susceptibility to *P. expansum*, as well as to another wound-infecting species, *B. cinerea*, is apparently associated with ripening time; later cultivars appear to be more tolerant (Davey et al., 2007; Ahmadi-Afzadi et al., 2013; Tahir et al., 2015a). One reason could be that late-ripening apple cultivars generally have a lower ethylene-regulated climacteric burst and therefore higher fruit firmness.

7 Genetic determination of texture

Increasingly accurate genetic maps for apples are now available, with candidate genes that have an impact on ethylene production. In 2012, numerous QTLs

Table 1 List of apple texture loci mentioned in this chapter, their linkage groups, major effects reported and references

Name	LG	Effect	Ref.
Md-Exp7	1	cell wall breakdown: softening	Costa et al., 2008
QTL mealiness	1	mealiness	Kunihisa et al., 2016
Md-ERF4	3	ethylene regulation: firmness and softening	Hu et al., 2020
Md-ERF3	3	ethylene regulation: softening	Wu et al., 2021
NAC18.1	3	regulates ripening: firmness and softening	Migicovsky et al., 2021
Md-ACO1	10	ethylene regulation: firmness and softening	Costa et al., 2005
Md-PG1	10	cell wall breakdown: firmness/ crispness, softening, juiciness	Longhi et al., 2013; Poles et al., 2020
QTL watercore	14	watercore	Kunihisa et al., 2016
Md-ACS1	15	ethylene regulation: firmness and softening	Oraguzie et al., 2004
Md-ACS3a	15	ethylene regulation: softening	Wang et al., 2009
Md-ERF118	16	ethylene regulation: softening	Wu et al., 2021
Ma	16	malic acid regulation: acidity but also firmness	Ru et al., 2021
Md-XTH	16	cellulose/hemicellulose network: crispness retention	Chang and Tong, 2020
Md-β-Gal	?	ethylene regulation: fruit ripening	Farneti et al., 2021

were identified as having a significant impact on fruit texture (firmness, hardness, crispness, juiciness, sponginess, compression and slow breakdown) and correlated to 10 of the 17 LGs in apple (review in Marondedze and Thomas, 2013). Associations between allelic configuration in various candidate genes and fruit texture parameters have been shown, as these genes also co-occur with several of the previously identified QTLs (Costa et al., 2010; Longhi et al., 2012) (Table 1).

Md-ACS1 (1-aminocyclopropane-1-carboxylate synthase) has a significant influence on fruit firmness and rate of softening during storage (Oraguzie et al., 2004). Two alleles were identified: allele 2 is associated with reduced ethylene production and thus firmer fruit and slower softening while allele 1 results in normal ethylene production. The three alleles found in *Md-ACS3a* also appear to be closely associated with ethylene production and shelf-life in apples (Wang et al., 2009). Another gene that can affect fruit firmness is the biallelic *Md-ACO1* (1-aminocyclopropane-1-carboxylate oxidase; Costa et al., 2005, 2010; Zhu and Barritt, 2008). The *Md-ACS1* and *Md-ACO1* genes have been mapped to LG 15 and LG 10, respectively (Costa et al., 2005; Chagné et al., 2019) and DNA markers have been applied to the screening of several large sets of apple cultivars (Sunako et al., 1999; Harada et al., 2000; Oraguzie et al., 2004, 2007; Nybom et al., 2008; Zhu and Barritt, 2008).

Interestingly, QTLs for different texture traits, as well as related traits such as ripening period, are often located in the same regions, suggesting that genetic variation in one or several interrelated genes may affect numerous traits simultaneously. Some of the steps in the ethylene-driven regulatory network determining fruit firmness and softening were revealed recently in a gene expression study following the detection of a mutation in the *ETHYLENE RESPONSE FACTOR (ERF4)* gene on LG 3 (Hu et al., 2020).

Much research has also focused on the cell wall and its degradation during fruit ripening. The biallelic endopolygalacturonase gene *Md-PG1* has been mapped to LG 10, just 37 cm from *Md-ACO1* (Costa et al., 2010; Longhi et al., 2013; Chagné et al., 2019). Allelic variation in *Md-PG1* can explain up to 40% of the phenotyped variation in firmness and texture, and DNA markers are now being exploited in several apple breeding programs around the world. Genetic configuration can, together with fruit flesh anatomy, explain some of the phenotypic variations in other texture traits; homozygosity for the low-softening allele in *Md-PG1*, in combination with a high fraction of rounded cells in the fruit flesh, was found to be strongly associated with high levels of juiciness in a set of 14 apple cultivars and selections (Poles et al., 2020). Heterozygous cultivars and/or cultivars with more angular cell shapes showed lower levels of juiciness.

Expansin enzymes affect depolymerization of different polysaccharides in the cell walls during fruit softening and may therefore also be of interest. An expansin homolog, *Md-Exp7*, was mapped on LG 1, and a functional marker was developed by Costa et al. (2008). However, screening a large set of genotypes has shown that there are multiple alleles, and their effects are difficult to interpret (Nybom et al., 2013).

In contrast to the instrumentally derived phenotypic data generally used in research on fruit texture genetics, Amyotte et al. (2017) performed a GWAS (genome-wide association study) based on descriptive data provided by a trained sensory panel for a collection of 85 apple cultivars. Genomic associations were recorded for several traits including crispness, juiciness and mealiness. Some of the detected QTLs resided in genomic regions not implicated in previous studies and may therefore target different genes.

The availability of large numbers of mapped single nucleotide polymorphisms (SNPs) has simplified the detection of QTLs and determination of minor gene impact on texture traits. Firmness, which can be scored rapidly in large materials, has now been investigated in traditional biparental mapping populations as well as in pedigree-based analyses (PBA) and GWAS. In one study, a GWAS was performed on a collection of 233 apple accessions

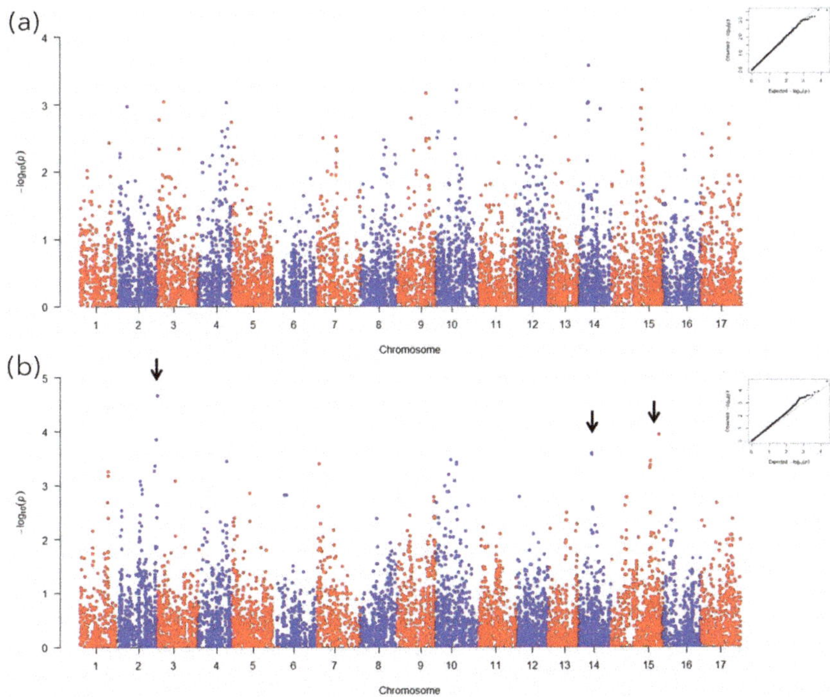

Figure 3 Manhattan plot illustrating the association between SNP markers and two texture subtraits assessed with a high-resolution texture analyzer, maximum force (a) and number of acoustic peaks (b), computed in a panel of 233 apple accessions. The *x*- and *y*-axes report the number of chromosomes and the -log10(*P*-value), respectively. For both panels, the Q-Q plot is also reported. Source: Reproduced with permission from Di Guardo et al. (2017).

together with a PBA of six full-sib pedigreed families (Di Guardo et al., 2017). DNA polymorphisms were determined with a 20K SNP array and fruit texture assessments with a sophisticated high-resolution texture analyzer. QTLs were identified on LG 10 (especially mechanical properties) and LG 2 and 14 (acoustic response properties) (Fig. 3).

In another study, more than 8000 SNPs were used to develop a model for genomic selection in apples (Roth et al., 2020). A set of 537 genotypes were first phenotyped for mechanical and acoustic parameters, and a training set of 259 genotypes with high phenotypic variability was chosen. A principal components analysis showed that the two major components correspond to firmness and crispness, respectively, with firmness being the most accurately estimated. Although most attention has focused on firmness and crispness, QTLs for other traits have also been detected, e.g. one QTL for mealiness in LG 1 in a set of 'Fuji'-related accessions (Kunihisa et al., 2016).

Recently, transcriptomic analyses have revealed additional genes involved in ethylene production and cell wall structure and degradation, which are differentially expressed in genotypes with opposing fruit texture traits. Several candidate genes for crispness retention during storage were revealed in a study of 'Honeycrisp' and some of its offspring, including *Md-XTH*, which regulates the cell wall-modifying enzyme xyloglucan endotransglucosylase/hydrolase (Chang and Tong, 2020). Another interesting gene is *Md-β-Gal2*, which determines β-galactosidase levels and appears to be associated with fruit ripening (Yang et al., 2018).

In addition to genes directly involved in ethylene biosynthesis and pectin degradation, several transcription factors such as ethylene response factors (*ERF*) which bind to, for example, the *Md-ACS1* promoter, can have a significant impact on fruit texture parameters. Wu et al. (2021) used bulked segregant analysis and RNA-seq to identify 62 QTLs in a set of 2664 apple seedlings. A total of 56 candidate genes were analyzed and shown to predict 55% and 60% of the variation in retention of firmness and crispness, respectively. Functional validation of these candidates provided evidence that small deletions in the *Md-ERF3* promoter (LG 3) and in the *Md-ERF118* promoter (LG 16) reduce the retention of firmness and crispness following storage through their impact on several important genes.

Another transcription factor gene, *NAC18.1*, affected both firmness at harvest and softening during storage in 800 apple accessions (Migicovsky et al., 2021). This gene, located on LG 3, outperformed the commonly screened *Md-ACS1*, *Md-ACO1* and *Md-PG1* in terms of predicting firmness at harvest and after storage. Softening rate was, however, more accurately predicted with *Md-PG1*. The *NAC18.1* gene encodes a protein that is orthologous to a transcription factor (*NON-RIPENING*), which regulates ripening in tomatoes. It can be presumed that it acts by modulating the ripening process in apples as well.

8 Association between texture and aroma

Apple flavor can be defined as a combination of taste (perception in the mouth) and aroma (mainly perceived by olfactory receptors). Taste is often estimated as the amount of sugars and organic acids, and the ratio between these compounds. Aroma is dependent on a complex cultivar-specific mixture of many volatile organic compounds (VOCs), estimated to comprise at least 300 different molecules in apples (Ulrich and Dunemann, 2012). The delimitation between these concepts is not exact since fruity esters can contribute to the taste of sweetness (Ting et al., 2015; Aprea et al., 2017).

The aroma profile changes as the fruit ripens, starting with a predominance of aldehydes, followed by an increase in alcohols and ending with a predominance of ester compounds at maturity (Fellman et al., 2000).

Consequently, ripening-related processes play an important role in determining the type and amount of aroma in any given fruit. Aroma is also affected by environmental factors, as shown in a set of 11 cultivars grown in two different locations in northern Italy (Chitarrini et al., 2020). Some currently applied storage methods, as well as treatment with 1-MCP, can alter the VOCs as well as the sensory perception of the fruit in a negative way (Ting et al., 2015).

Unfortunately, cultivars with favorable fruit texture (firm and crispy) generally have a comparatively restricted aromatic profile, while highly aromatic cultivars tend to be softer. Fruits of 162 cultivars grown in northern Italy were analyzed for texture and aroma as well as VOC phenotyping following 2 months of cold storage (Farneti et al., 2017). An SNP-based GWAS indicated that *Md-PG1* is not only responsible for fruit texture but also has a direct effect on the quantitative emission of volatile compounds. A gene expression analysis similarly showed that *Md-PG1*, together with four other texture genes (*Md-ACS1, Md-ACS3, Md-*β*-Gal* and *Md-ACO1*), was up-regulated toward the end of a 2-month storage period of the high-ethylene 'Golden Delicious' but remained at a basal level in the almost non-climacteric 'Fuji'. However, a burst of ethylene is necessary to stimulate the activation of genes fundamental to the production of volatile compounds (Busatto et al., 2016).

The physiological disorder watercore is caused by the rapid breakdown of cell walls and fluid filling up the intercellular spaces. The commonly occurring watercore fruits of several old European apple cultivars used to be appreciated for their enhanced sweetness and juiciness but are now avoided due to flesh break-down and browning during long-term storage. Watercore is, however, still desirable in some Asian countries where affected fruit is marketed as being sweeter and more aromatic. Apparently, ethyl ester synthesis is enhanced under hypoxic conditions within watercored tissues, resulting in a distinctive, fermented flavor (Tanaka et al., 2020). While many watercore-prone cultivars rapidly become soft and mealy, other cultivars like 'Fuji' remain firm in spite of developing watercore. A QTL for watercore development in 'Fuji' and its relatives has been detected in LG 14 (Kunihisa et al., 2016). Although watercore and mealiness often occur together, many newly developed watercore-susceptible lines derived from 'Fuji' lack this mealiness and can have both excellent flavor and texture (Tanaka et al., 2020).

9 Selection/breeding achievements

Selective sweeps of several traits have helped shape modern apples, as deduced from genetic sequences of modern apple cultivars and progenitor species such as *Malus sieversii* and *M. silvestris* (Duan et al., 2017). One region on LG 16 with several polygalacturonases, one region on LG 17 with cellulose synthase genes, and another on LG 12 with pectin esterases have experienced intensive human selection for fruit firmness. Further examination of these regions may assist in detecting genes for utilization in modern apple breeding.

A more recent development has involved the selection of particular genes, although the heritage of the targeted traits was unknown. The frequency of the firmness-promoting allele 2 of *Md-ACS1* has thus increased substantially, from below 20% in older cultivars (developed before 1800) to more than 50% in recently registered cultivars (from 1960 onward), indicating that this allele has been favored by selection for improved fruit quality in modern apple breeding programs (Nybom et al., 2008). Allele frequencies for numerous genes were similarly quantified in ancestral accessions and progenies included in the apple REFPOP (Jung et al., 2022). Compared to ancestral accessions, the allele with an increasing effect on phenotype had a higher frequency in the progeny for fruit firmness as well as for some other traits (later harvest date, increased flowering intensity, titratable acidity and trunk increment).

To date, thousands of apple cultivars have been selected and propagated around the world, and prominent gene flow across Europe has been revealed (Urrestarazu et al., 2016). Commercial production is, however, dominated by a very small number of genotypes, most of which are closely related (Muranty et al., 2020), and access to highly variable germplasm is therefore crucial for breeding purposes. Moreover, the need for carefully defined protocols for sensory and instrumental data collection has been highlighted in large phenotyping projects involving unreplicated genotypes planted at different locations (Schmitz et al., 2013).

New cultivars are generally developed by crossing suitable parental genotypes followed by screening the resulting seedlings in the field and evaluating propagated trees in test orchards. Large-scale screenings of textural traits undertaken in apple progenies have shown there is ample room for further improvement. Histological phenotyping revealed substantial variation in cell size distribution in a biparental apple progeny while sensory and instrumental analyses revealed variation in parameters, including firmness, crispness, graininess and juiciness (Gálvez-López et al., 2011a), as well as chemical composition and structure of cell wall polysaccharides (Gálvez-López et al., 2011b).

For breeding and selection purposes, the heritability of the different fruit texture parameters in a relevant plant material is crucial. Several estimates around 50% have been reported for firmness but crispness is usually considerably lower, e.g. 23% in a study on apple seedlings from 25 biparental crosses (Ru et al., 2021). Many estimations have, however, been based on mapping populations where very different genotypes are contrasted and may not be fully valid for apple breeding programs focusing on elite genotypes.

Recently, MAB (marker-assisted breeding) and MAS (marker-assisted selection) have been implemented in some apple breeding programs, both for major genes and for QTLs. The RosBREED SNP Consortium OpenArray v.1.0

assay with 128 SNPs linked to loci that determine fruit quality and pest and disease resistance was developed for MAS in breeding programs (Chagné et al., 2019). This array includes 15 SNPs across five fruit texture trait loci (*Md-PG1* and *Md-ACO1* on LG 10, *Md-ACS1* and QTLs derived from 'Braeburn' on LG 15 and LG 16). Three *Md-PG1* SNPs and one SNP for *Md-ACS1* proved to be significantly associated with BLUP (Best Linear Unbiased Prediction) values for firmness and crispness of advanced selections and commercial cultivars and were further validated using average phenotypes in advanced selection and validation families.

DNA markers are often used to screen putative parents. The cost-effectiveness of screening the seedlings depends on the percentage of seedlings that can be discarded at an early stage (Wannemuehler et al., 2019), which in turn depends on the number and efficiency of the DNA markers scored. When 127 Swedish-grown apple cultivars were screened for allelic composition in four fruit texture genes, alleles previously described as having good texture were associated with significantly lower softening for *Md-ACS1* and *Md-PG1*, but the opposite was noted for *Md-EXP7*, while results were insignificant for *Md-ACO1* (Nybom et al., 2013). These markers accounted for 15% of the observed variation in initial firmness and 18% for softening rate. The inclusion of ripening period, storage time (i.e. 6 or 12 weeks) and initial firmness into the model increased the predictability of softening rate to 38%.

In another study based on 321 seedlings in a US breeding program, genotyping with *Md-ACS1*, *Md-ACO1*, *Md-PG1* and *Md-Ma* (mainly affecting acidity) produced a 16% increase in additive variance for apple crispness and 17% for firmness (Ru et al., 2021). Of the different loci tested, only *Md-PG1* was significant for both firmness and crispness while *Md-Ma* was significant for firmness but still explained much less than *Md-PG1*. Similarly, only *Md-PG1*, together with the newly detected *NAC18.1*, had any predictive power in a set of 800 cultivars phenotyped for firmness at harvest and after storage and softening (Migicovsky et al., 2021). Even when the ripening period (harvest date) was entered into the model, variation in *NAC18.1* could only predict 18% of the variation in firmness. More powerful markers are therefore needed for large-scale, cost-effective applications in breeding programs.

10 Conclusion and future trends

Of several described texture traits, firmness, crispness, juiciness and mealiness are the most studied, with data obtained from both instrumental and sensory analyses. Fruit anatomy (cell shape and size) and physiology (ethylene production and enzymatic degradation of cell walls) have a major impact on fruit texture, as do environmental factors such as altitude and orchard management practices. Fruit texture is an important determinant for storability of different cultivars as well as their tolerance to fungal storage diseases, but unfortunately there appears to be a negative relationship between flavor and firmness.

Several major genes and QTLs have been identified as having a significant impact on firmness, and DNA markers have been developed for use in plant breeding. More efficient markers for firmness as well as for other texture traits are, however, needed for seedling selection with a high predictability of good fruit texture.

In the future, large genotyping arrays, such as the 20K and 480K SNP developed for apple (Bianco et al., 2014, 2016) and large-scale genotyping-by-sequencing, together with improved data analyses such as GWAS and GAP (Genomics-Assisted Prediction) and transcriptomics will improve our understanding of the intricate networks that regulate fruit texture traits. Development of DNA assays with carefully selected markers for key genes will become increasingly important in applied apple breeding programs, provided that cost-effective levels of selection success can be achieved.

Lack of high-quality phenotyping data has been identified as a critical factor. In addition to exact and informative measurements, the potential to assess genotype-environment interaction by measuring the same trait over several years - and preferably on trees growing in different orchards - would be very helpful. Recently, large field collections have been implemented, such as REFPOP where the same apple genotype is planted at six different locations (Jung et al., 2020). Although data is not yet available for all locations, close co-localization between markers for harvest date and fruit firmness has already been shown on LG 16 while firmness and several other traits co-occurred with firmness on LG 3 (Jung et al. 2022).

11 Where to look for further information

11.1 Further reading

Bejaei et al. (2021) present a recent analysis of relationships between sensory and instrumental texture trait data.

Chagné et al. (2019) describe the development and validation of a SNP array for MAS.

Musacchi and Serra (2018) provide a review of pre-harvest factors that have an impact on apple fruit quality. Part of this overview is dedicated to fruit texture but it should also be read for insights into other aspects of fruit quality, many of which are interrelated with texture.

Nybom et al. (2020) describe the impact of apple fruit ripening, texture and chemical contents on genetically determined susceptibility to storage rots.

Published by Burleigh Dodds Science Publishing Limited, 2024.

Wannemuehler et al. (2019) analyze the cost-effectiveness of applying MAB and MAS in apple breeding programs.

Wu et al. (2021) provide a detailed analysis of the regulatory network involved in retention of firmness and crispness during storage.

11.2 Key conferences

EUCARPIA (European Association for Research on Plant Breeding), Section Fruit, organizes conferences every 3–4 years that are relevant to fruit breeding and genetics.

ISHS (International Society for Horticultural Science) organizes a world congress every 4 years, as well as symposia on narrower topics such as biotechnology and molecular breeding, fruit production and post harvest.

RGC (International Rosaceae Genomics Conference) is organized every 2 years and coordinated by the Rosaceae International Genomics Initiative.

11.3 Major research centers and international projects

Multi-partner projects, such as the EU-funded FP7 Fruitbreedomics (2010–2015) and the USDA-SCRI funded RosBREED 1 and RosBREED 2 projects (2010–2018) (www.rosbreed.org) have played a large role in developing international research on apple genetics. Both have resulted in several spin-off projects, emphasizing the need for international cooperation.

In addition to the numerous European and American research facilities involved in the abovementioned projects, valuable research is carried out by Summerland Research and Development Centre and Vineland Research and Innovation Centre in Canada, The New Zealand Institute for Plant and Food Research, and China Agricultural University in Beijing.

12 References

Ahmadi-Afzadi, M., Tahir, I. and Nybom, H. (2013). Impact of harvesting time and fruit firmness on the tolerance to fungal storage diseases in an apple germplasm collection, *Postharvest Biol. Technol.* 82: 51–58.

Amyotte, B., Bowen, A. J., Banks, T., Rajcan, I. and Somers, D. J. (2017). Mapping the sensory perception of apple using descriptive sensory evaluation in a genome wide association study, *PLoS ONE* 12(2): e0171710, https://doi.org/10.1371/journal.pone.0171710.

Aprea, E., Charles, M., Endrizzi, I., Corollaro, M. L., Betta, E., Biasioli, F. and Gasperi, F. (2017). Sweet taste in apple: The role of sorbitol, individual sugars, organic acids and volatile compounds, *Sci. Rep.* 7: 44950, https://https://doi.org/10.1038/srep44950.

Bejaei, M. (2022). Converting apple textural parameters obtained from penetrometers and their relationships with sensory attributes, *Horticulturae* 8(3): 269, https://doi.org/10.3390/horticulturae8030269.

Bejaei, M., Stanich, K. and Cliff, M. A. (2021). Modelling and classification of apple textural attributes using sensory, instrumental and compositional analyses, *Foods* 10(2): 384, https://doi.org/10.3390/foods10020384.

Bianco, L., Cestaro, A., Linsmith, G., Muranty, H., Denancé, C., Théron, A., Poncet, C., Micheletti, D., Kerschbamer, E., Di Pierro, E. A., Larger, S., Pindo, M., van de Weg, E., Davassi, A., Laurens, F., Velasco, R., Durel, C. E. and Troggio, M. (2016). Development and validation of the Axiom® Apple480K SNP genotyping array, *Plant J.* 86(1): 62–74.

Bianco, L., Cestaro, A., Sargent, D. J., Banchi, E., Derdak, S., Di Guardo, M., Salvi, S., Jansen, J., Viola, R., Gut, I., Laurens, F., Chagné, D., Velasco, R., van de Weg, E. and Troggio, M. (2014). Development and validation of a 20K single nucleotide polymorphism (SNP) whole genome genotyping array for apple (*Malus* × *domestica* Borkh.), *PLoS ONE* 9(10): e110377, https://doi.org/10.1371/journal.pone.0110377.

Biggs, A. R. and Miller, S. S. (2001). Relative susceptibility of selected apple cultivars to *Colletotrichum acutatum*, *Plant Dis.* 85(6): 657–660.

Blažek, J., Opatová, H., Goliáš, J. and Homutová, I. (2007). Ideotype of apples with resistance to storage diseases, *Hort. Sci. (Prague)* 34(3): 107–113.

Brummell, D. A. and Harpster, M. H. (2001). Cell wall metabolism in fruit softening and quality and its manipulation in transgenic plants, *Plant Mol. Biol.* 47(1–2): 311–340.

Buccheri, M. and Di Vaio, C. (2005). Relationship among seed number, quality, and calcium content in apple fruits, *J. Plant Nutr.* 27(10): 1735–1746.

Busatto, N., Farneti, B., Tadiello, A., Velasco, R., Costa, G. and Costa, F. (2016). Candidate gene expression profile reveals a time specific activation among different harvest times in 'Golden Delicious' and 'Fuji' apple cultivars, *Euphytica* 208(2): 401–413, https://doi.org/10.1007/s10681-015-1621-y.

Butkeviciute, A., Viskelis, J., Viskelis, P., Liaudanskas, M. and Janulis, V. (2021). Changes in the biochemical composition and physicochemical properties of apples stored in controlled atmosphere conditions, *Appl. Sci.* 11(13): 6215, https://doi.org/10.3390/app11136215.

Chagné, D., Vanderzande, S., Kirk, C., Profitt, N., Weskett, R., Gardiner, S. E., Peace, C. P., Volz, R. K. and Bassil, N. V. (2019). Validation of SNP markers for fruit quality and disease resistance loci in apple (*Malus* × *domestica* Borkh.) using the OpenArray® platform, *Hortic. Res.* 6: 30, https://doi.org/10.1038/s41438-018-0114-2.

Chang, H. Y. and Tong, C. B. S. (2020). Identification of candidate genes involved in fruit ripening and crispness retention through transcriptome analyses of a 'Honeycrisp' population, *Plants (Basel)* 9(10): 1335, https://doi.org/10.3390/plants9101335.

Charles, M., Corollaro, M. L., Manfrini, L., Endrizzi, I., Aprea, E., Zanella, A., Grappadelli, L. C. and Gasperi, F. (2018). Application of sensory-instrumental tool to study apple texture characteristics shaped by altitude and time of harvest, *J. Sci. Food Agric.* 98(3): 1095–1104.

Chitarrini, G., Dordevic, N., Guerra, W., Robatscher, P. and Lozano, L. (2020). Aroma investigation of new and standard apple varieties grown at two altitudes using gas chromatography-mass spectrometry combined with sensory analysis, *Molecules* 25(13): 3007, https://doi.org/10.3390/molecules25133007.

Corollaro, M. L., Aprea, E., Endrizzi, I., Betta, E., Demattè, M. L., Charles, M., Bergamaschi, M., Costa, F., Biasioli, F., Corelli Grappadelli, L. and Gasperi, F. (2014). A combined sensory-instrumental tool for apple quality evaluation, *Postharvest Biol. Technol.* 96: 135–144.

Costa, F. (2016). Mechanical investigation to assess the peel contribution in apple fruit, *Postharvest Biol. Technol.* 111: 41–47.

Costa, F., Cappellin, L., Fontanari, M., Longhi, S., Guerra, W., Magnago, P., Gasperi, F. and Biasioli, F. (2012). Texture dynamics during postharvest cold storage ripening in apple (*Malus × domestica* Borkh.), *Postharvest Biol. Technol.* 69: 54–63.

Costa, F., Cappellin, L., Longhi, S., Guerra, W., Magnago, P., Porro, D., Soukoulis, C., Salvi, S., Velasco, R., Biasioli, F. and Gasperi, F. (2011). Assessment of apple (*Malus × domestica* Borkh.) fruit texture by a combined acoustic-mechanical profiling strategy, *Postharvest Biol. Technol.* 61(1): 21–28.

Costa, F., Peace, C. P., Stella, S., Serra, S., Musacchi, S., Bazzani, M., Sansavini, S. and Van de Weg, W. E. (2010). QTL dynamics for fruit firmness and softening around an ethylene-dependent polygalacturonase gene in apple (*Malus × domestica* Borkh.), *J. Exp. Bot.* 61(11): 3029–3039.

Costa, F., Stella, S., Van de Weg, W. E., Guerra, W., Cecchinel, M., Dalla Via, J., Koller, B. and Sansavini, S. (2005). Role of the genes *Md-ACO1* and *Md-ACS1* in ethylene production and shelf life of apple (*Malus domestica* Borkh), *Euphytica* 141(1–2): 181–190.

Costa, F., Van de Weg, W. E., Stella, S., Dondini, L., Pratesi, D., Musacchi, S. and Sansavini, S. (2008). Map position and functional allelic diversity of *Md-Exp7*, a new putative expansin gene associated with fruit softening in apple (*Malus × domestica* Borkh.) and pear (*Pyrus communis*), *Tree Genet. Genomes* 4(3): 575–586.

Daillant-Spinnler, B., MacFie, H., Beyts, P. and Hedderley, D. (1996). Relationships between perceived sensory properties and major preference directions of 12 varieties of apples from the southern hemisphere, *Food Qual. Pref.* 7: 113–126.

Davey, M. W., Auwerkerken, A. and Keulemans, J. (2007). Relation of apple vitamin C and antioxidant contents to harvest date and postharvest pathogen infection, *J. Sci. Food Agric.* 87(5): 802–813.

Di Guardo, M., Bink, M. C. A. M., Guerra, W., Letschka, T., Lozano, L., Busatto, N., Poles, L., Tadiello, A., Bianco, L., Visser, R. G. F., van de Weg, E. and Costa, F. (2017). Deciphering the genetic control of fruit texture in apple by multiple-family based analysis and genome-wide association, *J. Exp. Bot.* 68(7): 1451–1466.

Duan, N., Bai, Y., Sun, H., Wang, N., Ma, Y., Li, M., Wang, X., Jiao, C., Legall, N., Mao, L., Wan, S., Wang, K., He, T., Feng, S., Zhang, Z., Mao, Z., Shen, X., Chen, X., Jiang, Y., Wu, S., Yin, C., Ge, S., Yang, L., Jiang, S., Xu, H., Liu, J., Wang, D., Qu, C., Wang, Y., Zuo, W., Xiang, L., Liu, C., Zhang, D., Gao, Y., Xu, Y., Xu, K., Chao, T., Fazio, G., Shu, H. and Zhong, G. (2017). Genome re-sequencing reveals the history of apple and supports a two-stage model for fruit enlargement, *Nat. Commun.* X: 8, 249, https://doi.org/10.1038/s41467-017-00336-7.

Farneti, B., Di Guardo, M., Khomenko, I., Capellin, L., Biasioli, F., Velasco, R. and Costa, F. (2017). Genome-wide association study unravels the genetic control of the apple volatilome and its interplay with fruit texture, *J. Exp. Bot.* 68(7): 1467–1478.

Fellman, J. K., Miller, T. W., Mattinson, D. S. and Mattheis, J. P. (2000). Factors that influence biosynthesis of volatile flavor compounds in apple fruits, *Hortscience* 35(6): 1026–1033.

Gálvez-López, D., Laurens, F., Devaux, M. F. and Lahaye, M. (2012). Texture analysis in an apple progeny through instrumental, sensory and histological phenotyping, *Euphytica* 185(2): 171–183.

Gálvez-López, D., Laurens, F., Quéméner, B. and Lahaye, M. (2011). Variability of cell wall polysaccharides composition and hemicellulose enzymatic profile in an apple progeny, *Int. J. Biol. Macromol.* 49(5): 1104–1109.

Grotte, M., Duprat, F., Loonis, D. and Piétri, E. (2001). Mechanical properties of the skin and the flesh of apples, *Int. J. Food Prop.* 4(1): 149–161.

Gwanpua, S. G., Dakwa, V., Verboven, P., Nicolai, B. M., Geeraerd, A. F., Hendrickx, M., Christiaens, S. and Verlinden, B. E. (2015). Relationship between texture analysis and texture attributes during postharvest softening of 'Jonagold' and 'Kanzi' apples, *Acta Horticult.* 1079: 279–284.

Gwanpua, S. G., Verlinden, B. E., Hertog, M. L. A. T. M., Nicolai, B. M., Hendrickx, M. and Geeraerd, A. H. (2018). Understanding the regulation of texture degradation during apple softening – a kinetic approach, *Acta Horticult.* 1194: 196, https://doi.org/10.17660/ActaHortic.2018.1194.196.

Hampson, C. R., Quamme, H. A., Hall, J. W., MacDonald, R. A., King,M. C. and Cliff, M. A. (2000). Sensory evaluation as a selection tool in apple breeding, *Euphytica* 111(2): 79–90.

Harada, T., Sunako, T., Wakasa, Y., Soejima, J., Satoh, T. and Niizeki, M. (2000). An allele of the 1-aminocyclopropane-1-carboxylate synthase gene (*Md-ACS1*) accounts for the low level of ethylene production in climacteric fruits of some apple cultivars, *Theor. Appl. Genet.* 101(5–6): 742–746.

Hu, Y., Han, Z., Sun, Y., Wang, S., Wang, T., Wang, Y., Xu, K., Zhang, X., Xu, X., Han, Z. and Wu, T. (2020). ERF4 affects fruit firmness through TPL4 by reducing ethylene production, *Plant J.* 103(3): 937–950.

Janisiewicz, W. J., Saftner, R. A., Conway, W. S. and Forsline, P. L. (2008). Preliminary evaluation of apple germplasm from Kazakhstan for resistance to postharvest blue mold in fruit caused by *Penicillium expansum*, *Hortscience* 43(2): 420–426.

Johnston, J. W., Hewett, E. W., Hertog, M. and Harker, F. R. (2002). Temperature and ethylene affect induction of rapid softening in 'Granny Smith' and 'Pacific Rose' apple cultivars, *Postharvest Biol. Technol.* 25(3): 257–264.

Jönsson, Å. and Nybom, H. (2007). Consumer evaluation of scab-resistant apple cultivars in Sweden, *Agric. Food Sci.* 15(4): 388–401.

Jung, M., Keller, B., Roth, M., Aranzana, M. J., Auwerkerken, A., Guerra, W., Al-Rifai, M., Lewandowski, M., Sanin, N., Rymenants, M., Didelot, F. and Dujak, C. (2022). Genetic architecture and genomic predictive ability of apple quantitative traits across environments, *Horticult. Res.* 9, uhac028, https://doi.org/10.1093/hr/uhac028.

Jung, M., Roth, M., Aranzana, M. J., Auwerkerken, A., Bink, M., Denancé, C., Dujak, C. and Durel, C.-E. (2020). The apple REFPOP—A reference population for genomics-assisted breeding in apple, *Horticult. Res.* 7: 189, https://doi.org/10.1038/s41438-020-00408-8.

Kim, M. S., Duizer, L. M. and Grygorczyk, A. (2022). Application of a texture analyzer friction rig to evaluate complex texture attributes in apples, *Postharvest Biol. Technol.* 186: 111820, https://doi.org/10.1016/j.postharvbio.2021.111820.

Kunihisa, M., Moriya, S., Abe, K., Okada, K., Haji, T., Hayashi, T., Kawahara, Y., Itoh, R., Itoh, T., Katayose, Y., Kanamori, H., Matsumoto, T., Mori, S., Sasaki, H., Matsumoto, T., Nishitani, C., Terakami, S. and Yamamoto, T. (2016). Genomic dissection of a 'Fuji' apple cultivar: Re-sequencing, SNP marker development, definition of haplotypes, and QTL detection, *Breed. Sci.* 66(4): 499–515.

Li, H., Liu, J., Zhang, X., Zhu, Z., Yang, H., Dang, M. and Zhao, Z. (2019). Comparison of textural and ultrastructural characteristics of four apple cultivars with different textures during cold storage, *Int. J. Food Prop.* 22(1): 659–669.

Longhi, S., Hamblin, M. T., Trainotti, L., Peace, C. P., Velasco, R. and Costa, F. (2013). A candidate gene based approach validates *Md-PG1* as the main responsible for a QTL impacting fruit texture in apple (*Malus* × *domestica* Borkh), *BMC Plant Biol.* 13: 37, www.biomedcentral.com/1471-2229/13/37.

Longhi, S., Moretto, M., Viola, R.,Velasco, R. and Costa, F. (2012). Comprehensive QTL mapping survey dissects the complex fruit texture physiology in apple (*Malus* × *domestica* Borkh.), *J. Exp. Bot.* 63(3): 1107-1121.

Mann, H., Bedford, D., Luby, J., Vickers, Z. and Tong, C. (2005). Relationship of instrumental and sensory texture measurements of fresh and stored apples to cell number and size, *HortSci.* 40(6): 1815–1820.

Marondedze, C. and Thomas, L. (2013). Genes and quality trait loci (QTLs) associated with firmness in *Malus* × *domestica*, *Afr. J. Biotechnol.* 12: 996–1003.

Migicovsky, Z., Yeats, T. H., Watts, S., Song, J., Forney, C. F., Burgher-MacLellan, K., Somers, D. J., Gong, Y., Zhang, Z., Vrebalov, J., van Velzen, R., Giovannoni, J. G., Rose, J. K. C. and Myles, S. (2021). Apple ripening is controlled by a NAC transcription factor, *Front. Genet.* 12: 671300, https://doi.org/10.3389/fgene.2021.671300.

Mpelasoka, B. S., Behboudian, M. H. and Mills, T. M. (2001). Effects of deficit irrigation on fruit maturity and quality of 'Braeburn' apple, *Sci. Horticult.* 90(3–4): 279–290.

Muranty, H., Denancé, C., Feugey, L., Crépin, J. L., Barbier, Y., Tartarini, S., Ordidge, M., Troggio, M., Lateur, M., Nybom, H., Paprstein, F., Laurens, F. and Durel, C. E. (2020). Using whole-genome SNP data to reconstruct a large multi-generation pedigree in apple germplasm, *BMC Plant Biol.* 20(1): 2, https://doi.org/10.1186/s12870-019-2171-6.

Musacchi, S. and Serra, S. (2018). Apple fruit quality: Overview on pre-harvest factors, *Sci. Horticult* 234: 409–430.

Naeem-Abadi, T., Keshavarzi, M., Alaee, H., Hajnagari, H. and Hoseinava, S. (2014). Blue mold (*Penicillium expansum*) decay resistance in apple cultivars, and its association with fruit physicochemical traits, *J. Agric. Sci. Technol.* 16: 635–644.

Ng, J. K. T., Schröder, R., Brummell, D. A., Sutherland, P. W., Hallett, I. C., Smith, B. G., Melton, L. D. and Johnston, J. W. (2015). Lower cell wall pectin solubilisation and galactose loss during early fruit development in apple (*Malus* × *domestica*) cultivar 'Scifresh' are associated with slower softening rate, *J. Plant Physiol.* 176: 129-137.

Norelli, J. L., Wisniewski, M., Fazio, G., Burchard, E., Gutierrez, B., Levin, E. and Droby, S. (2017). Genotyping-by-sequencing markers facilitate the identification of quantitative trait loci controlling resistance to *Penicillium expansum* in *Malus sieversii*, *PLoS ONE* 12(3): e0172949, https://doi.org/10.1371/journal.pone.0172949.

Nybom, H., Ahmadi-Afzadi, M., Rumpunen, K. and Tahir, I. (2020). Review of the impact of apple fruit ripening, texture and chemical contents on genetically determined susceptibility to storage rots, *Plants (Basel)* 9(7): 831, https://doi.org/10.3390/plants9070831.

Nybom, H., Ahmadi-Afzadi, M., Sehic, J. and Hertog, M. (2013). DNA marker-assisted evaluation of fruit firmness at harvest and post-harvest fruit softening in a diverse apple germplasm, *Tree Genet. Genomes* 9(1): 279–290.

Nybom, H., Sehic, J. and Garkava-Gustavsson, L. (2008). Modern apple breeding is associated with a significant change in allelic ratio of the ethylene production gene *Md-ACS1*, *J. Hortic. Sci. Biotechnol.* 83(5): 673–677.

Oraguzie, N. C., Iwanami, H., Soejima, J., Harada, T. and Hall, A. (2004). Inheritance of the *Md-ACS1* gene and its relationship to fruit softening in apple (*Malus* × *domestica* Borkh.), *Theor. Appl. Genet.* 108(8): 1526–1533.

Oraguzie, N. C., Volz, R. K., Whitworth, C. J., Bassett, H. C. M., Hall, A. J. and Gardiner, S. E. (2007). Influence of *Md-ACS1* allelotype and harvest season within an apple germplasm collection on fruit softening during cold air storage, *Postharvest Biol. Technol.* 44(3): 212–219.

Peck, G. M., Andrews, P. K., Reganold, J. P. and Fellman, J. K. (2006). Apple orchard productivity and fruit quality under organic, conventional and integrated management, *Hortscience* 41(1): 99–107.

Piazza, L. and Giovenzana, V. (2015). Instrumental acoustic-mechanical measures of crispness in apples, *Food Res. Intl.* 69: 209–215.

Poles, L., Gentile, A., Giuffrida, A., Valentini, L., Endrizzi, I., Aprea, E., Gasperi, F., Distefano, G., Artioli, G., La Malfa, A., Costa, F., Lovatti, L. and Di Guardo, M. (2020). Role of fruit flesh morphology and *MdPG1* allelotype in influencing juiciness and texture properties in apple, *Postharvest Biol. Technol.* 164, 111161, https://doi.org/10.1016/j.postharvbio.2020.111161.

Reganold, J. P., Glover, J. D., Andrews, P. K. and Hinman, H. R. (2001). Sustainability of three apple production systems, *Nature* 410(6831): 926–930.

Roth, M., Muranty, H., Di Guardo, M., Guerra, W., Patocchi, A. and Costa, F. (2020). Genomic prediction of fruit texture and training population optimization towards the application of genomic selection in apple, *Hortic. Res.* 7(148): 148, https://doi.org/10.1038/s41438-020-00370-5.

Ru, S., Hardner, C., Evans, K., Main, D., Carter, P. A., Harshman, J., Sandefur, P., Edge-Garza, D. and Peace, C. (2021). Empirical evaluation of multi-trait DNA testing in an apple seedling population, *Tree Genet. Genomes* 17(1): 13, https://doi.org/10.1007/s11295-021-01494-y.

Sadar, N., Agati, G. and Zanella, A. (2018). Optical, acoustic and textural attributes in 'Braeburn' and 'Nicoter' (Kanzi®) apple resulting from different pre- and postharvest conditions, *Acta Hortic.* (1194): 753–760.

Schmitz, C. A., Clark, M. D., Luby, J. J., Bradeen, J. M., Guan, Y., Evans, K., Orcheski, B., Brown, S., Verma, S. and Peace, C. (2013). Fruit texture phenotypes of the RosBREED U.S. apple reference germplasm set, *Hortscience* 48(3): 296–303.

Serra, S., Leisso, R., Giordani, L., Kalcsits, L. and Musacchi, S. (2016). Crop load influences fruit quality, nutritional balance, and return bloom in 'Honeycrisp' apple, *Hortscience* 51(3): 236–244.

Spoor, T., Rumpunen, K., Sehic, J., Ekholm, A., Tahir, I. and Nybom, H. (2019). Chemical contents and blue mould susceptibility in Swedish-grown cider apple cultivars, *Eur. J. Horticult Sci.* 84(3): 131–141.

Sunako, T., Sakuraba, W., Senda, M., Akada, S., Ishikawa, R., Niizeki, M. and Harada, T. (1999). An allele of the ripening-specific 1-aminocyclopropane-1-carboxylic acid synthase gene (*ACS1*) in apple fruit with a long storage life, *Plant Physiol.* 119(4): 1297–1304.

Published by Burleigh Dodds Science Publishing Limited, 2024.

Tahir, I. I. and Gustavsson, K. E. (2010). Improving quality and storability of apples by a combination of aluminum reflective mulch, summer pruning and controlled nitrogen fertilization, *Acta Hortic.* (877): 245–250.

Tahir, I. I. and Nybom, H. (2013). Tailoring organic apples by cultivar selection, production system, and postharvest treatment to improve quality and storage life, *Hortscience* 48(1): 92–101.

Tahir, I. I., Nybom, H., Ahmadi-Afzadi, M., Røen, K., Sehic, J. and Røen, D. (2015a). Susceptibility to blue mold caused by *Penicillium expansum* in apple cultivars adapted to a cool climate, *Eur. J. Hortic. Sci.* 80(3): 117–127.

Tahir, I. I., Svensson, S.-E. and Hansson, D. (2015b). Floor management systems in an organic apple orchard affect fruit quality and storage life, *Hort.Sci.* 50(3): 434–441.

Tanaka, F., Hayakawa, F. and Tatsuki, M. (2020). Flavor and texture characteristics of 'Fuji' and related app (*Malus domestica* L.) cultivars, focusing on the rich watercore, *Molecules* 25(5): 1114, https://doi.org/10.3390/molecules25051114.

Teh, S. L., Brutcher, L., Schonberg, B. and Evans, K. (2020). Eleven-year correlation of physical fruit texture traits between computerized penetrometers and sensory assessment in an apple breeding program, *hortTechnology* 30(6): 719–724.

Tijskens, L. M. M., Schouten, R. E., Zanella, A. and Sadar, N. (2018). Apples from Monalisa – Biological variation of firmness behaviour in storage and shelf life, *Acta Horticult* 1194: 1415–1420.

Ting, V. J. L., Romano, A., Silcock, P., Bremer, P. J., Corollaro, M. L., Soukoulis, C., Capellin, L., Gasperi, F. and Biasioli, F. (2015). Apple flavor: Linking sensory perception to volatile release and textural properties, *J. Sens. Stud.* 30(3): 195–210, https://doi.org/10.1111/joss.12151.

Ulrich, D. and Dunemann, F. (2012). Towards the development of molecular markers for apple volatiles, *Flavour Fragr. J.* 27(4): 286–289.

Urrestarazu, J., Denancé, C., Ravon, E., Guyader, A., Guisnel, R., Feugey, L., Poncet, C., Lateur, M., Houben, P., Ordidge, M., Fernandez-Fernandez, F., Evans, K. M., Paprstein, F., Sedlak, J., Nybom, H., Garkava-Gustavsson, L., Miranda, C., Gassmann, J., Kellerhalls, M., Suprun, I., Pikunova, A. V., Krasova, N. G., Torutaeva, E., Dondini, L., Tartarini, S., Laurens, F. and Durel, C. E. (2016). Analysis of the genetic diversity and structure across a wide range of germplasm reveals prominent gene flow in apple at the European level, *BMC Plant Biol.* 16(1): 130. https://doi.org/10.1186/s12870-016-0818-0.

Vanoli, M., Grassi, M., Buccheri, M., Lovati, F., Sadar, N., Zanella, A., Torricelli, A., Rizzolo, A. and Spinelli, L. (2018a). Time-resolved reflectance spectroscopy reveals different texture characteristics in 'Brabeurn', 'Gala' and 'Kanzi®' apples, *Acta Horticult.* 1194: 1273–1282.

Vanoli, M., Lovati, F., Grassi, M., Buccheri, M., Zanella, A., Cattaneo, T. M. P. and Rissolo, A. (2018b). Water spectral pattern as a marker for studying apple sensory texture, *Adv. Horticult Sci.* 32: 343–351.

Vigneau, E., Charles, M. and Chen, M. (2014). External preference segmentation with additional information on consumers: A case study on apples, *Food Qual. Pref.* 32: 83–92.

Wakasa, Y., Kudo, H., Ishikawa, R., Akada, S., Senda, M., Niizeki, M. and Harada, T. (2006). Low expression of an endopolygalacturonase gene in apple fruit with long-term storage potential, *Postharvest Biol. Technol.* 39(2): 193–198.

Wang, A., Yamakake, J., Kudo, H., Wakasa, Y., Hatsuyama, Y., Igarashi, M., Kasai, A., Li, T. Z. and Harada, T. (2009). Null mutation of the *MdACS3* gene, coding for a ripening-specific 1-aminocyclopropane-1-carboxylate synthase, leads to long shelf life in apple fruit, *Plant Physiol.* 151(1): 391–399.

Wannemuehler, S. D., Luby, J. J., Yue, C., Bedford, D. S., Gallardo, R. K. and McCracken, V. A. (2019). A cost–benefit analysis of DNA informed apple breeding, *Hortscience* 54(11): 1998–2004, https://doi.org/10.21273/HORTSCI14173-19.

Wei, J. M., Ma, F. W., Shi, S. G., Qi, X. D., Zhu, X. Q. and Yuan, J. W. (2010). Changes and postharvest regulation of activity and gene expression of enzymes related to cell wall degradation in ripening apple fruit, *Postharvest Biol. Technol.* 56(2): 147–154.

Win, N. M., Yoo, J., Naing, A. H., Kwon, J.-G. and Kang, I.-K. (2021). 1-methylcyclopropene (1-MCP) treatment delays modification of cell wall pectin and fruit softening in "Hwangok" and "Picnic" apples during cold storage, *Postharvest Biol. Technol.* 180, https://doi.org/10.1016/j.postharvbio.2021.111599.

Wu, B., Shen, F., Wang, X., Zheng, W. Y., Xiao, C., Deng, Y., Wang, T., Yu Huang, Z., Zhou, Q., Wang, Y., Wu, T., Feng Xu, X., Hai Han, Z. and Zhong Zhang, X. (2021). Role of *MdERF3* and *MdERF118* natural variations in apple flesh firmness/crispness retainability and development of QTL-based genomics-assisted prediction, *Plant Biotechnol. J.* 19(5): 1022–1037.

Yang, H., Liu, J., Dang, M., Zhang, B., Li, H., Meng, R., Qu, D., Yang, Y. and Zhao, Z. (2018). Analysis of β-galactosidase during fruit development and ripening in two different texture types of apple cultivars, *Front. Plant Sci.* 9: 539, https://doi.org/10.3389/fpls.2018.00539.

Yang, L., Cong, P., He, J., Bu, H., Qin, S. and Lyu, D. (2022). Differential pulp cell wall structures lead to diverse fruit textures in apple (*Malus domestica*), *Protoplasma* 259(5): 1205–1217.

Zdunek, A., Cybulska, J., Konopacka, D. and Rutkowski, K. (2011). Evaluation of apple texture with contact acoustic emission detector: A study on performance of calibration models, *J. Food Eng.* 106(1): 80–87.

Zhu, Y. and Barritt, B. H. (2008). *Md-ACS1* and *Md-ACO1* genotyping of apple (*Malus × domestica* Borkh.) breeding parents and suitability for marker-assisted selection, *Tree Genet. Genomes* 4(3): 555–562.

Chapter 3

Advances in understanding the nutritional and nutraceutical properties of apples

Gabriela Ploscuțanu, "Dunărea de Jos" University of Galați, Romania

1 Introduction

Global consumption of fruits is increasing due to their nutraceutical role in protecting human health from the development of chronic disease. Hosseini et al. (2018) e.g. suggested that a higher intake of fruits and vegetables could lead to both a reduction in pro-inflammatory mediators and an improved immune cell profile. Fruits, in general, are therefore promoted as healthy products. Over the last few decades, breeding and management shaped the biodiversity of fruits towards our dietary preferences, which have evolved together with the availability of different fruits in different agro-ecological regions.

The use and maintenance of this bio-cultural heritage in contemporary food systems depend on a wide range of social, cultural, political, environmental and economic factors, and there is an important interplay between the current and future availability of fruit biodiversity, and current and future dietary diversity (Harris et al., 2022). The already high number of research studies suggesting a correlation between fruit consumption and the reduced risk of major chronic diseases has continued to rise. There are thousands of fruit species from which to choose but, despite this abundance, many consumers still do not consume the recommended daily amounts of fruits.

There is huge potential to better incorporate the wealth of diverse fruit species and varieties into food systems (Kennedy et al., 2021). In 2003, the

http://dx.doi.org/10.19103/AS.2023.0127.05

World Health Organization (WHO) recommended the consumption of at least 400 g of fruits and vegetables per day – the equivalent of five daily portions. Prompted by WHO recommendations, there is increasing interest in promoting fruits and vegetables for better health and wellbeing. As a result, public and private partnerships have developed in many countries to educate and inform consumers regarding recommended daily dosages of fruit.

A variety of factors influence consumption and the barriers that might limit their intake. Several worldwide campaigns have been carried out, with varying levels of success in terms of raising awareness and increasing the consumption of fruits and vegetables (Rekhy and McConchie, 2014). These campaigns provide international guidelines recommending the consumption of two servings of fruits and three servings of vegetables per day (Miller et al., 2016). The United Nations declared 2021 as the International Year of Fruits and Vegetables in order to raise awareness of the nutritional and health benefits of consuming more fruits and vegetables as part of a diversified, balanced, and healthy diet, alongside a healthy lifestyle. Additionally, the UN also highlighted the need to reduce the loss and waste of these highly perishable items (FAO, 2020).

However, products high in fat, salt, and sugar are advertised intensively on social media and media platforms, including digital platforms used daily by children, and can be purchased almost everywhere. These advertising campaigns contribute to the high rate of metabolic disorders, including obesity and diabetes. The majority of children and adolescents do not consume the recommended amount of fruits and vegetables. Reducing the advertising of energy-dense snacks and increasing the marketing of healthier foods, such as fruits, is a necessary strategy to improve the dietary intake of children and to decrease the risk of chronic metabolic disorders in later life (Folkvord et al., 2021).

Phytochemicals are found in fruits and vegetables and protect plants against bacteria, viruses and fungi. Consuming large amounts of brightly coloured fruits and vegetables (yellow, orange, red, green, white, blue, purple), whole grains/cereals and beans containing phytochemicals can reduce the risk of developing disorders such as metabolic diseases, cardiovascular diseases (CVDs) or cancer. The action of phytochemicals varies with their colour and source. They may act as antioxidants in preventing carcinogens (cancer-causing agents) from forming (Cosme et al., 2022).

Phenolic compounds are bioactives that act as antioxidants by reacting with a variety of free radicals. The mechanism of antioxidant actions involves either hydrogen atom transfer or transfer of a single electron or sequential proton loss electron transfer or by chelation of transition metals (Zeb, 2020). In fruits, polyphenols have gained particular attention due to their association with both *in vitro* and *in vivo* improved antioxidant activity related to their ability to scavenge free radicals (Preti and Tarola, 2021). This mechanism allows them to play a major role in the prevention of chronic diseases (Penny et al., 2017). Fruits provide several important compounds, including dietary fibre, which are linked to a lower incidence of CVD and obesity. Fruits are also important sources of vitamins, minerals and other biologically active phytochemicals that act as phytoestrogens, antioxidant and anti-inflammatory agents.

The apple (*Malus domestica* Borkh) is one of the most popular fruits consumed worldwide. Its success is linked to its flavour and taste, year-round availability and nutritional properties. The phytochemical compounds and the anti-inflammatory nutrients (polyphenols, anthocyanins, flavonoids, etc.) present in apples have many beneficial effects in consumer health (Jaglan et al., 2021). Apples can prevent the risk of chronic disorders through various mechanisms, including antioxidant, anti-inflammatory, antiproliferative and cell-signalling effects. Apples and their derived products have also been associated with beneficial effects on the risk and markers of neoplasms, CVDs, asthma and Alzheimer's disease. Recent research has suggested that these products may also be associated with improved outcomes in regard to cognitive decline associated with aging, diabetes, weight management, bone health, lung function and gastrointestinal protection (Hyson, 2011).

2 The phytochemical composition of apples

Apple (*M. domestica)* has a high concentration of polyphenols (Bohn and Bouayed, 2020; Feng et al., 2021). The composition and content of biologically active compounds in apples differ according to tissue types and cultivars. In addition to their high pectin, vitamin and mineral content, they are also an excellent source of antioxidants capable of capturing and neutralising free radicals which, in turn, play a role in preventing occurrence of CVDs and neoplasms.

The concentration of its phytochemicals varies with variety and post-harvest storage conditions. The phytochemical profile of apples differs greatly between varieties, and there are also some changes in phytochemicals during fruit ripening (Boyer and Liu, 2004). Environmental factors such as climate, soil condition, place of cultivation and post-harvesting storage conditions also have an influence on the overall content of the fruit. Many environmental factors influence the accumulation of polyphenols in apples, such as exposure to ultraviolet light, climatic conditions and soil conditions, such as nitrogen supply (Rupasinghe et al., 2012). Temperature and light parameters also contribute to changes in the external and internal quality of the apple, such as red overcolour and the accumulation of dry matter. Another parameter is crop loading and thinning – processes that can induce physiological changes that can, in turn, lead to a higher content of dry matter. Irrigation can also alter colouring of the skin and the chemical composition of the pulp (Musacchi and Serra, 2018). Factors including light, temperature, mineral nutrition, growth regulators and carbohydrate availability can affect pigment concentration in the skin (Dar et al., 2019). The ripening process can also change the composition of apples (OECD, 2019). Several studies have indicated that storage has little to no effect on phytochemical content, but (as discussed in the next section) processing can greatly affect the concentration of apple bioactives (Boyer and Liu, 2004).

The main sensory attributes of the apple (responsible for sweetness, acidity or bitterness) are provided by components such as sugars, organic acids and polyphenolic compounds. Polyphenols are distributed differently throughout the fruit (peel or pulp) (Biedrzycka and Amarowicz, 2008). The typical composition of apple is summarised in Table 1. In addition to simple carbohydrates (mainly various monosaccharides and disaccharides), apples contain a high amount of vitamin C, minerals (especially potassium), as well as several other beneficial constituents such as dietary fibre and phytochemicals such as phytosterols and triterpenes (Bohn and Bouayed, 2020).

Polyphenols represent the major class of antioxidants in apples (Fig. 1). A large variety of these components have been associated with their anti-inflammatory and antioxidant mechanisms in epidemiological studies, by decreasing the risk of chronic diseases such as diabetes and cardiovascular disorders (Bohn and Bouayed, 2020).

The polyphenolic compounds present in apples can be divided into two major groups:

- Flavonoids – flavones, flavonols, flavanols, flavanones, isoflavones, proanthocyanidins, anthocyanins and their derivatives; and
- Phenolic acids – hydroxybenzoic acids and hydroxycinnamic acids.

Flavonoids present in apples are divided into different classes:

- Flavonols such as quercetin and its glycosylated forms (3-galactoside, 3-glucoside, 3-rhamnoside);
- Flavan-3-ols derivatives including monomers (catechin and epicatechin); and
- Dimers, trimers and other condensed tannins known as procyanidins.

Apples have been found to contain flavan-3-ols in significant amounts (Fabbrini et al., 2022). Apple procyanidins are a group of oligomeric and polymeric flavonoids formed through the association of flavan-3-ol units epicatechin and catechin (Maldonado-Celis et al., 2009; Kalinowska et al., 2014). The broad-spectrum antimicrobial properties of quercetin e.g. have been suggested as an alternative to antibiotics in preventing disease (Yang et al., 2020).

Table 1 Composition of apple fruits with respect to macro- and micronutrients as well as major phytochemical classes (Bohn and Bouayed, 2020)

Constituent	Amount	Reference(s)
Water (g/100 g)	85	Souci et al. (2000) and USDA (2015)
Lipids (g/100 g)	0.2–0.6	
Proteins (g/100 g)	0.25–0.45	
Carbohydrates (g/100 g)	10.4--11.5	
• Fructose	2.0	Souci et al. (2000)
• Glucose	5.7	
• Sucrose	2.5	
Dietary fibre (g/100 g)	2.0–2.4	Souci et al. (2000)
• Water soluble	0.5	
• Water insoluble	1.5	
• Magnesium	3–9	
• Calcium	4–11	
• Potassium	100–175	
• Sodium	1	
Phytosterols (mg/100 g)	12/30	Souci et al. (2000) and Rudell et al. (2011)
• Beta-sitosterol	11	
• Campesterol	1	
Total polyphenols (mg/100 g)	85–430[a]	Bouayed et al. (2011)
Total polyphenols (mg/100 g)	50[b]	Rothwell et al. (2013)
Total triterpenes (mg/100 g)	40–350	Jäger et al. (2009) and Andre et al. (2012)
Vitamin C (mg/100 g)	12	Souci et al. (2000)
Vitamin E (mg/100 g)	0.2–0.5	Souci et al. (2000) and USDA (2015)
Total carotenoids (μg/100 g)	37–46	Souci et al. (2000)

[a] Determined via Folin–Ciocalteu (bearing the risk of overestimation).
[b] Sum of individual polyphenols determined by HPLC-UPLC.

In terms of phenolic acids, apples have a high concentration of several hydroxybenzoic acids such as *p*-hydroxybenzoic acid, protocatechuic acid, gallic acid, syringic acid, gentisic acid and several hydroxycinnamic acids such as *p*-coumaric acid, caffeic acid, ferulic acid or chlorogenic acid. Several dihydrochalcones such as phlorizin and its derivatives and anthocyanidins such as cyanidins and their glycosides have also been identified (Bohn and Bouayed, 2020).

Several studies have been undertaken to identify the phytochemical composition of different varieties of apples. Wojdyło et al. (2008) studied the phenolic composition of 67 apple cultivars (new and old varieties) in order to assess the concentration of the most important phytochemicals. The average

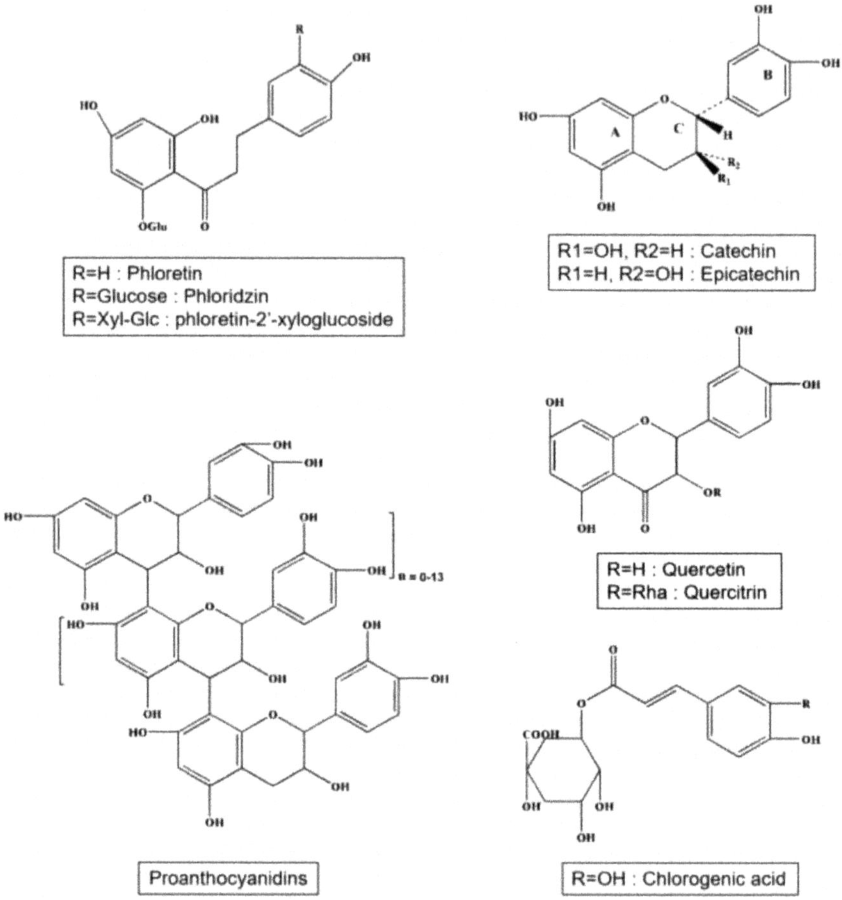

Figure 1 Structures of main phenolic acids, dihydrochalcones and flavonoids in apple. Source: From Patocka et al. (2020).

content of total polyphenols was between 523 mg/100 g DW and 2724.96 mg/100 g DW, depending on the apple variety.

Bondonno et al. (2017) found that flavanols (catechin and oligomeric procyanidins) represented the major class of polyphenols in apples, representing more than 80%, followed by hydroxycinnamic acids (1–31%), flavonols (2–10%), dihydrochalcones (0.5–5%) and, in red apples, anthocyanins (1%). Other important compounds found in apples include chlorogenic acid, (+)-catechin, (−)-epicatechin, procyanidin B2 dimer, phlorizin, cyanidin-3-O-galactoside, quercetin-3-O-galactoside, quercetin-5-caffeoylquinic acid, etc.

Another study involved the phytochemical composition of organic apples from several certified organic and conventional orchards located in central Poland (Mazovia region). These were compared to conventionally cultivated apples, with organic products characterised by significantly higher concentrations of phenolic acids (av. >31%) and flavonols (av. >66%) (Średnicka-Tober et al., 2020). Vitamin C content was strongly dependent on year-to-year differences in growing conditions. The study suggested that the organic production system had the potential to provide fruits richer in health-promoting phenolic antioxidants (Średnicka-Tober et al., 2020).

Butkevičiūtė et al. (2022) studied the composition of apple peel and determined that the highest total triterpenes' content was approximately 8.0 mg/g. Depending on the rootstock, apple peel accumulated between 3.52 and 4.74 times more triterpenes compounds than the concentration found in apple pulp samples. Ursolic acid was the main triterpene compound identified in both the peel and pulp samples (Butkevičiūtė et al., 2022).

Rupasinghe et al. (2012) assessed the phytochemical composition of several varieties of apples. The distribution and concentration of polyphenols varies greatly within apples and apple cultivars. The highest concentration was found for phloridzin, amongst the dihydrochalcones group. The peel had a higher content of polyphenols than the pulp or core, being abundant in flavonoids such as quercetin glycosides and cyanidine galactosides. The pulp and the core pith presented relatively high concentrations of chlorogenic acid (Rupasinghe et al., 2012).

Vasile et al. (2021) investigated red-skinned apple extracts of the Generos variety from Romania. The samples revealed the presence of cyanidin-3-O-galactoside, cyanidin-3-O-glucoside and cyanidin-3-O-arabinoside. Bongiorni et al. (2022) examined the Tuscia Red apple variety from Italy in terms of anthocyanin content and identified several compounds, mainly cyanidin-derived compounds. Based on the findings, it is recommended that apples are consumed complete with peel which contains more bioactive compounds than the pulp (Boyer and Liu, 2004).

3 Nutraceuticals in apple products

Salazar-Orbea et al. (2023) studied the effects of freezing, heat treatment, high-pressure processing as well as storage up to 12 months on composition. Hot or cold crushing processes to produce apple puree had a greater impact than storage. Proanthocyanidins was the major group found in apple puree and the most stable amongst the phenolic group during storage, while anthocyanins were most affected by both processing and storage. Flavonols and dihydrochalcones were also quite stable (Salazar-Orbea et al., 2023). Feng et al. (2021) studied both apples and apple cider and concluded that the most abundant phytochemicals groups were phenolic acids and flavan-3-ols.

In terms of canned products, the deconstruction of the fruits' fibrous matrix by thermal and mechanical treatments, combined with the addition of simple types of sugars, degraded the nutritional quality of the fruits by lowering their nutritional density and a matrix effect. New 'non-thermal' processes (e.g. pulsed electric fields, high-pressure, supercritical CO_2, radiation, etc.) appear promising as they limit the losses of important compounds such as vitamin C and antioxidant phytonutrients while allowing a satisfactory storage time. To prolong the preservation time of the products, drying may cause loss of antioxidants (Fardet and Richonnet, 2020).

Studies have suggested that the consumption of apple juice has positive effects on the markers related to CVD, neoplasms and neurodegenerative diseases (Mossine et al., 2020; Vallée Marcotte et al., 2022).. The beneficial effects of apple juice are more obvious in the cloudy variety than the clear, probably due to a higher nutrient density in the former. Li et al. (2022) assessed the impact of titanium dioxide nanoparticles (TiO_2 NPs) on the composition and antioxidant activity of apple juice polyphenols. The results showed that an addition of 500 mg/100 g TiO_2 NPs significantly decreased the concentration of total polyphenols, attributed to the formation of polyphenols-TiO_2 NPs charge transfer complexes.

Le Bourvellec et al. (2011) assessed the composition of apple sauces and found that these products contain high concentrations of phytochemicals such as flavonols and dihydrochalcones. An oxidation product of dihydrochalcone was also detected. The concentration of hydroxycinnamic acids and flavan-3-ols decreased proportionally depending on the variety with high levels of flavan-3-ols.

There is growing interest in making use of by-products from processing fruits such as apples to support a more circular, green bioeconomy (Fierascu et al., 2020). Apple by-products, mainly pomace, contain insoluble sugars, including cellulose (127.9 g/kg dry weight (DW)), hemicellulose (7.2–43.6 g/kg DW), lignin (15.3–23.5 g/kg DW) and simple carbohydrates such as glucose, fructose and galactose. Apple pomace is an important source of antioxidant compounds, in particular quercetin glycosides, epicatechin, phloridzin, phloretin, chlorogenic acid and other polyphenolic constituents (Lu and Yeap Foo, 2000). Apple pomace can be used in various food systems following minimal processing or in the form of extracts, significantly increasing the functional value of food and contributing to the reduction of food waste (Lyu et al., 2020). Phloridzin is an important biological marker for apple pomace.

Apple pomace is a potentially valuable source of numerous bioactive compounds, such as polyphenols, in the development of functional food products and nutraceuticals (Pollini et al., 2021). The encapsulation of bioactive compounds is another promising alternative for the use of agro-industrial by-products as active ingredients in the food industry. Several studies have shown that bioactives are extractable from by-products, and their stability and solubility can be improved by using appropriate encapsulation techniques, to provide antioxidant, antimicrobial and anti-inflammatory properties (Marcillo-Parra et al., 2021).

4 The role of apple nutraceutical compounds in promoting health and preventing disease

Chronic non-communicable disorders (e.g. cardiovascular disorders, neurodegenerative diseases, neoplasms, diabetes, obesity) are characterised by meta-inflammation which involves a metabolic cascade, including cellular oxidative stress, atherosclerotic processes and insulin resistance which gradually causes significant damage to the body (Koch, 2019). Nutrition and diet are seen as essential ways to prevent this process and decrease the rate of modern cardio-metabolic diseases, including obesity, metabolic syndrome, type 2 diabetes and atherosclerosis (Nasri et al., 2014). In particular, the crosstalk between polyphenols and gut microbiota, recently revealed thanks to DNA-based tools and next-generation sequencing, has revealed the central regulatory role of dietary polyphenols and their gut microecological metabolites on host energy metabolism and associated diseases (Koudoufio et al., 2020; Carbajal et al., 2022; Awuchi and Okpala, 2022).

A diet rich in polyphenols protects against many chronic pathologies by modulating numerous physiological processes such as cellular redox potential, enzymes activity, cell proliferation and signal transduction pathways. However, polyphenols possess a low oral bioavailability, mainly due to the extensive biotransformation mediated by the phase I and phase II reactions that occur in enterocytes and the liver, but also by gut microbiota. However, most polyphenols have proven significant biological effects that highlight the low bioavailability/high bioactivity paradox (Fotirić Akšić et al., 2022). The term bioavailability involves several variables, such as intestinal absorption, excretion of glucuronides to the intestinal lumen, microflora metabolism, intestinal and hepatic metabolism, plasma kinetics, nature of circulating metabolites, albumin binding, cellular uptake, intracellular metabolism, tissue accumulation and biliary and urinary excretion. Some polyphenols may be absorbed less efficiently than others, but still reach equivalent plasma concentrations due to lower secretion into the intestinal lumen and lower metabolism and elimination (Manach et al., 2004). The processes of digestion and absorption as generally accepted today for polyphenols are illustrated in Fig. 2.

A distinction must be made between the absorption of low- and high-molecular-weight polyphenols along the digestive tract. In terms of absorption kinetics, low-molecular-weight polyphenols have the highest bioavailability. These include isoflavones, caffeic and gallic acids lead, catechins, flavanones

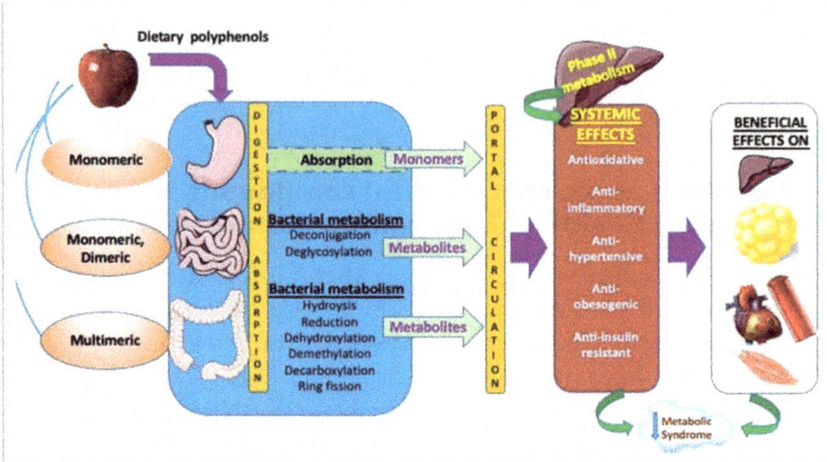

Figure 2 Integrated view of intestinal dietary polyphenol absorption, luminal transformation, and actions of their relevant metabolites on cardiometabolic disorders. Source: Adapted from: Koudoufio et al. (2020).

and quercetin glucosides. High-molecular-weight polyphenols such as proanthocyanidins, galloylated catechins and anthocyanins are less bioavailable (Manach et al., 2004; Koudoufio et al., 2020).

Compounds in apples, particularly phenolic compounds, have been found to possess high antioxidant activity, inhibit cancer cell proliferation, decrease lipid oxidation and lower cholesterol and reduce onset of diabetes (Aprikian et al., 2003; Boyer and Liu, 2004; Hodgson et al., 2015; Pucci et al., 2017; Šamec et al., 2021).

A consistent inverse association has been demonstrated between apple consumption and the risk of various types of cancer, including colorectal cancer (Jedrychowski and Maugeri, 2009; Bracci et al., 2021). Flavan-3-ols have been found to protect against various types of cancer, including rectal, oropharyngeal, laryngeal, breast and stomach cancer (Lei et al., 2016).

Flavonoids, particularly some of their microbially derived metabolites (of e.g. enterolactone), may help reduce the development of CVD and the risk of CVD mortality (Wang et al., 2014). This evidence is supported by animal studies reporting the beneficial effects of polyphenols or pure compounds on CVD risk factors such as blood cholesterol, blood pressure, endothelial function and arterial stiffness (Del Rio et al., 2013). Apple consumption improves various CVD risk factors by lowering blood pressure, total cholesterol (TC), low-density lipoprotein (LDL) cholesterol (LDLc), while increasing high-density lipoprotein cholesterol (HDLc) and improving endothelial function (González-Sarrías et al., 2017; Sandoval-Ramírez et al., 2020). Studies have shown apple consumption to decrease the plasma concentrations of oxidised LDL/beta2-glycoprotein I complex (oxLDL-b2GPI), a targeted contributor for atherosclerosis (Zhao et al., 2013). Kim et al. (2022) determined that consumption of apples and apple-derived products for more than 1 week can reduce TC and LDL levels.

A significant number of studies have shown the antimicrobial activity of polyphenols against several pathogens, including herpes simplex virus, Epstein–Barr virus, enterovirus 71, influenza virus and other viruses that cause infections related to the respiratory tract (da Silva, 2021). Oligomeric procyanidins isolated from unripe apple peel have been shown to possess antiviral and immunostimulatory effects, including reducing Dengue virus infectivity (Kimmel et al., 2011).

Angiotensin-converting enzyme 2 (ACE2) represents a host receptor for the severe and acute respiratory syndrome Coronavirus 2 (SARS-CoV-2). A potential antiviral therapeutic approach is to inhibit the interaction between the top envelope glycoproteins (S proteins) of SARS-CoV-2 and ACE2. Polyphenols reduced rhACE2 activity by 15–66% at 10 μM. Rutin, quercetin-3-O-glucoside, tamarixetin and 3,4-dihydroxyphenylacetic acid inhibited rhACE2 activity by 42–48%. Quercetin was the most potent inhibitor of rhACE2 among the polyphenols tested, with an IC50 of 4.48 μM. Thus, quercetin, its metabolites and 3′,4′-hydroxylated polyphenols inhibited rhACE2 activity at physiologically relevant *in vitro* concentrations (Liu et al., 2020).

The three main apple polyphenols – namely chlorogenic acid, procyanidin dimer B2 and phloridzin – could bind to angiotensin-converting enzyme 2 (ACE2) antibodies *in vitro*, reducing their UV absorption intensity and fluorescence intensity. In addition, three apple polyphenols could modify the secondary structure of the antibodies and inhibit the recognition of ACE2 by its antibodies. Apple polyphenols may have some degree of *in vitro* inhibition on a class of proteins with important ACE2 recognition sites, so apple polyphenols may have the potential to inhibit the recognition of the viral proteins by ACE2, thereby preventing the virus from entering the cell host (Pu et al., 2022).

Numerous current clinical trials are investigating the effects of polyphenols in the prophylaxis and treatment of COVID-19, from symptomatic, through moderate and severe COVID-19 treatment, to antifibrotic treatment in discharged patients with COVID-19. The antiviral activities of polyphenols and their impact on the modulation of the immune system could serve as a solid basis for the development of natural polyphenol-based approaches for the prevention and treatment of COVID-19 (Gligorijevic et al., 2021).

Studies have also suggested that polyphenols in apples may help slow ageing processes such cognitive decline as well as help with diabetes, weight management, bone health, lung function and gastrointestinal function (Hyson, 2011). There have e.g. been a number of studies of the benefits of dried apple intake in weight reduction as well as reducing conditions hyperglycemia, which are recognised risk factors for cancer (Chai et al., 2012; Eisner et al., 2020; Mossine et al., 2020; Balderas et al., 2022). However, it is important to note that the efficacy of some nutraceuticals requires further study, including clinical control trials and randomised trials, especially for their bioactive composition and medicinal effects against specific disease conditions (Awuchi and Okpala, 2022).

5 Conclusion

Apples (*M. domestica*) are one of the oldest and most popular fruits in the world. While apples are mostly eaten fresh, they are also processed into beverages, jam, jelly and other forms of food. The phytochemical composition of apples is diverse. Polyphenols are the main bioactive and antioxidant compounds. Their stability, bio-accessibility, bioavailability and anti-inflammatory effects can be affected by many factors. Consumption of apple and its derived products or extracts has been associated with a reduced risk of developing neoplasms, cardiovascular disorders, diabetes and many other diseases. Promoting greater awareness of polyphenols and other bioactives found in apples could help to highlight the beneficial health effects of consuming this popular fruit amongst consumers. Apples also represent a valuable source for a range of nutraceutical products.

6 References

Andre, C. M., Greenwood, J. M., Walker, E. G., Rassam, M., Sullivan, M., Evers, D., Perry, N. B. and Laing, W. A. (2012). Anti-inflammatory procyanidins and triterpenes in 109 apple varieties. *Journal of Agricultural and Food Chemistry* 60(42), 10546–10554. https://doi.org/10.1021/jf302809k.

Aprikian, O., Duclos, V., Guyot, S., Besson, C., Manach, C., Bernalier, A., Morand, C., Rémésy, C. and Demigné, C. (2003). Apple pectin and a polyphenol-rich apple concentrate are more effective together than separately on cecal fermentations and plasma lipids in rats. *The Journal of Nutrition* 133(6), 1860–1865. https://doi.org/10.1093/jn/133.6.1860.

Awuchi, C. G. and Okpala, C. O. R. (2022). Natural nutraceuticals, especially functional foods, their major bioactive components, formulation, and health benefits for disease prevention: an overview. *Journal of Food Bioactives* 19, 97–123. https://doi.org/10.31665/JFB.2022.18317.

Balderas, C., Angulo, J., Sevilleja-Ortiz, A., Peiró, C., Vallejo, S., Dongil, P., de Ancos, B. and Sánchez-Moreno, C. (2022). Onion and apple functional ingredients intake improves antioxidant and inflammatory status and vascular injury in obese Zucker rats. *Antioxidants* 11(10), 1953–1982. http://doi.org/10.3390/antiox11101953.

Biedrzycka, E. and Amarowicz, R. (2008). Diet and health: apple polyphenols as antioxidants. *Food Reviews International* 24(2), 235–251. https://doi.org/10.1080/87559120801926302.

Bohn, T. and Bouayed, J. (2020). Apples: an apple a day, still keeping the doctor away? In: Jaiswal, A. K. (Ed.). *Nutritional Composition and Antioxidant Properties of Fruits and Vegetables*, Elsevier, Amsterdam, The Netherlands, pp. 595–612.

Bondonno, N. P., Bondonno, C. P., Ward, N. C., Hodgson, J. M. and Croft, K. D. (2017). The cardiovascular health benefits of apples: whole fruit vs. isolated compounds. *Trends in Food Science and Technology* 69, 243–256. https://doi.org/10.1016/j.tifs.2017.04.012.

Bongiorni, S., Arisi, I., Ceccantoni, B., Rossi, C., Cresta, C., Castellani, S., Forgione, I., Rinalducci, S., Muleo, R. and Prantera, G. (2022). Apple polyphenol diet extends lifespan, slows down mitotic rate and reduces morphometric parameters in Drosophila melanogaster: A comparison between three different apple cultivars. *Antioxidants* 11(11), 2086. https://doi.org/10.3390/antiox11112086.

Bouayed, J., Hoffmann, L. and Bohn, T. (2011). Antioxidative mechanisms of whole-apple antioxidants employing different varieties from Luxembourg. *Journal of Medicinal Food* 14(12), 1631–1637. https://doi.org/10.1089/jmf.2010.0260.

Boyer, J. and Liu, R. H. (2004). Apple phytochemicals and their health benefits. *Nutrition Journal* 3, 5. http://doi.org/10.1186/1475-2891-3-5.

Bracci, L., Fabbri, A., Del Cornò, M. and Conti, L. (2021). Dietary polyphenols: promising adjuvants for colorectal cancer therapies. *Cancers* 13(18), 4499–3516. https://doi.org/10.3390/cancers13184499.

Butkevičiūtė, A., Janulis, V. and Kviklys, D. (2022). Triterpene content in flesh and peel of apples Grown on different rootstocks. *Plants* 11(9), 1247. https://doi.org/10.3390/plants11091247.

Carbajal, I., Gutierrez-Chavez, D., Gonzalez-Arias, B., Jiménez-Hernández, A., Pacheco, I., Rico-García, E., Feregrino-Perez, A., Guevara-Gonzalez, R. and Ramirez Rodriguez, D. L. (2022). Biological macromolecules as nutraceuticals. In: Nayak, A. K., Dhara, A. K. and Pal, D. (Eds). *Biological Macromolecules. Bioactivity and Biomedical Applications*, Academic Press, London, pp. 97–138.

Chai, S. C., Hooshmand, S., Saadat, R. L., Payton, M. E., Brummel-Smith, K. and Arjmandi, B. H. (2012). Daily apple versus dried plum: impact on cardiovascular disease risk factors in postmenopausal women. *Journal of the Academy of Nutrition and Dietetics* 112(8), 1158–1168. https://doi.org/10.1016/j.jand.2012.05.005.

Cosme, F., Pinto, T., Aires, A., Morais, M. C., Bacelar, E., Anjos, R., Ferreira-Cardoso, J., Oliveira, I., Vilela, A. and Gonçalves, B. (2022). Red fruits composition and their health benefits - a review. *Foods* 11(5), 644. https://doi.org/10.3390/foods11050644.

Da Silva, A. P. G. (2021). Fighting coronaviruses with natural polyphenols. *Biocatalysis and Agricultural Biotechnology* 37, 102179. https://doi.org/10.1016/j.bcab.2021.102179.

Dar, J. A., Wani, A. A., Ahmed, M., Nazir, R., Zargar, S. M. and Javaid, K. (2019). Peel colour in apple (Malus × domestica Borkh.): an economic quality parameter in fruit market. *Scientia Horticulturae* 244, 50–60, https://doi.org/10.1016/j.scienta.2018.09.029.

Del Rio, D., Rodriguez-Mateos, A., Spencer, J. P., Tognolini, M., Borges, G. and Crozier, A. (2013). Dietary (poly)phenolics in human health: structures, bioavailability, and evidence of protective effects against chronic diseases. *Antioxidants and Redox Signaling* 18(14), 1818–1892. https://doi.org/10.1089/ars.2012.4581.

Eisner, A., Ramachandran, P., Cabalbag, C., Metti, D., Shamloufard, P., Kern, M., Hong, M. Y. and Hooshmand, S. (2020). Effects of dried apple consumption on body composition, serum lipid profile, glucose regulation, and inflammatory markers in overweight and obese children. *Journal of Medicinal Food* 23(3), 242–249. https://doi.org/10.1089/jmf.2019.0044.

Fabbrini, M., D'Amico, F., Barone, M., Conti, G., Mengoli, M., Brigidi, P. and Turroni, S. (2022). Polyphenol and tannin nutraceuticals and their metabolites: how the human gut microbiota influences their properties. *Biomolecules* 12(7), 875. https://doi.org/10.3390/biom12070875.

FAO (2020). Fruit and vegetables: your dietary essentials. In: *The International Year of Fruits and Vegetables, 2021*, background paper. Rome. https://doi.org/10.4060/cb2395en (accessed 15.03.2023).

Fardet, A. and Richonnet, C. (2020). Nutrient density and bioaccessibility, and the antioxidant, satiety, glycemic, and alkalinizing potentials of fruit-based foods according to the degree of processing: a narrative review. *Critical Reviews in Food Science and Nutrition* 60(19), 3233–3258. https://doi.org/10.1080/10408398.2019.1682512.

Feng, S., Yi, J., Li, X., Wu, X., Zhao, Y., Ma, Y. and Bi, J. (2021). Systematic review of phenolic compounds in apple fruits: compositions, distribution, absorption, metabolism, and processing stability. *Journal of Agricultural and Food Chemistry* 69(1), 7–27. https://doi.org/10.1021/acs.jafc.0c05481.

Fierascu, R. C., Sieniawska, E., Ortan, A., Fierascu, I. and Xiao, J. (2020). Fruits by-products: a source of valuable active principles. A short review. *Frontiers in Bioengineering and Biotechnology* 8, 319. https://doi.org/10.3389/fbioe.2020.00319.

Folkvord, F., Naderer, B., Coates, A. and Boyland, E. (2021). Promoting fruit and vegetable consumption for childhood obesity prevention. *Nutrients* 14(1), 157. https://doi.org/10.3390/nu14010157.

Fotirić Akšić, M., Nešović, M., Ćirić, I., Tešić, Ž., Pezo, L., Tosti, T., Gašić, U., Dojčinović, B., Lončar, B. and Meland, M. (2022). Polyphenolics and chemical profiles of domestic Norwegian apple (*Malus × domestica* Borkh.) cultivars. *Frontiers in Nutrition* 9, 941487. https://doi.org/10.3389/fnut.2022.941487.

Gligorijevic, N., Radomirovic, M., Nedic, O., Stojadinovic, M., Khulal, U., Stanic-Vucinic, D. and Cirkovic Velickovic, T. (2021). Molecular mechanisms of possible action of phenolic compounds in COVID-19 protection and prevention. *International Journal of Molecular Sciences* 22(22), 12385. http://doi.org/10.3390/ijms222212385.

González-Sarrías, A., Combet, E., Pinto, P., Mena, P., Dall'Asta, M., Garcia-Aloy, M., Rodríguez-Mateos, A., Gibney, E. R., Dumont, J., Massaro, M., Sánchez-Meca, J., Morand, C. and García-Conesa, M. T. (2017). A systematic review and meta-analysis of the effects of flavanol-containing tea, cocoa and apple products on body composition and blood lipids: exploring the factors responsible for variability in their efficacy. *Nutrients* 9(7), 746. https://doi.org/10.3390/nu9070746.

Harris, J. and Zonnevel, M. van, Achigan-Dako, E. G., Bajwa, B., Brouwer, I. D., Choudhury, D., Jager, I. de, Steenhuijsen Piters, B. de, Dulloo, M. E., Guarino, L., Kindt, R., Mayes, S., McMullin, S., Quintero, M. and Schreinemachers, P. (2022). Fruit and vegetable biodiversity for nutritionally diverse diets: Challenges, opportunities, and knowledge gaps. *Global Food Security*, 33, 100618. https://doi.org/10.1016/j.gfs.2022.100618.

Hodgson, K., Morris, J., Bridson, T., Govan, B., Rush, C. and Ketheesan, N. (2015). Immunological mechanisms contributing to the double burden of diabetes and intracellular bacterial infections. *Immunology* 144(2), 171–185. https://doi.org/10.1111/imm.12394.

Hosseini, B., Berthon, B. S., Saedisomeolia, A., Starkey, M. R., Collison, A., Wark, P. A. B. and Wood, L. G. (2018). Effects of fruit and vegetable consumption on inflammatory biomarkers and immune cell populations: a systematic literature review and meta-analysis. *The American Journal of Clinical Nutrition* 108(1), 136–155. https://doi.org/10.1093/ajcn/nqy082.

Hyson, D. A. (2011). A comprehensive review of apples and apple components and their relationship to human health. *Advances in Nutrition (Bethesda, Md.)* 2(5), 408–420. https://doi.org/10.3945/an.111.000513.

Jaglan, P., Buttar, H. S., Al-bawareed, O. A. and Chibisov, S. (2021). Potential health benefits of selected fruits: apples, blueberries, grapes, guavas, mangos, pomegranates, and tomatoes. In: Singh, R. B., Watanabe, S. and Isaza, A. A. (Eds). *Functional Foods and Nutraceuticals in Metabolic and Non-communicable Diseases*, Academic Press, San Diego, CA, pp. 359–370. https://doi.org/10.1016/B978-0-12-819815-5.00026-4.

Jäger, S., Trojan, H., Kopp, T., Laszczyk, M. N. and Scheffler, A. (2009). Pentacyclic triterpene distribution in various plants: rich sources for a new group of multi-potent plant extracts. *Molecules* 14(6), 2016–2031. https://doi.org/10.3390/molecules14062016.

Jedrychowski,, W. and Maugeri, U. (2009). An apple a day may hold colorectal cancer at bay: recent evidence from a case-control study. *Reviews on Environmental Health* 24(1), 59–74. https://doi.org/10.1515/REVEH.2009.24.1.59.

Kalinowska, M., Bielawska, A., Lewandowska-Siwkiewicz, H., Priebe, W. and Lewandowski, W. (2014). Apples: content of phenolic compounds vs. variety, part of apple and cultivation model, extraction of phenolic compounds, biological properties. *Plant Physiology and Biochemistry* 84, 169–188. https://doi.org/10.1016/j.plaphy.2014.09.006.

Kennedy, G., Kanter, R., Chotiboriboon, S., Covic, N., Delormier, T., Longvah, T., Maundu, P., Omidvar, N., Vish, P. and Kuhnlein, H. (2021). Traditional and indigenous fruits and vegetables for food system transformation. *Current Developments in Nutrition* 5(8), nzab092. https://doi.org/10.1093/cdn/nzab092.

Kim, S. J., Anh, N. H., Jung, C. W., Long, N. P., Park, S., Cho, Y. H., Yoon, Y. C., Lee, E. G., Kim, M., Son, E. Y., Kim, T. H., Deng, Y., Lim, J. and Kwon, S. W. (2022). Metabolic and cardiovascular benefits of apple and apple-derived products: a systematic review and meta-analysis of randomized controlled trials. *Frontiers in Nutrition* 9, 766155. https://doi.org/10.3389/fnut.2022.766155.

Kimmel, E. M., Jerome, M., Holderness, J., Snyder, D., Kemoli, S., Jutila, M. A. and Hedges, J. F. (2011). Oligomeric procyanidins stimulate innate antiviral immunity in dengue virus infected human PBMCS. *Antiviral Research* 90(1), 80–86. https://doi.org/10.1016/j.antiviral.2011.02.011.

Koch, W. (2019). Dietary polyphenols: important non-nutrients in the prevention of chronic noncommunicable diseases. A systematic review. *Nutrients* 11(5), 1039. http://doi.org/10.3390/nu11051039.

Koudoufio, M., Desjardins, Y., Feldman, F., Spahis, S., Delvin, E. and Levy, E. (2020). Insight into polyphenol and gut microbiota crosstalk: are their metabolites the key to understand protective effects against metabolic disorders? *Antioxidants*, 9(10), 982. http://doi.org/10.3390/antiox9100982.

Le Bourvellec, C., Bouzerzour, K., Ginies, C., Regis, S., Plé, Y. and Renard, C. M. G. C. (2011). Phenolic and polysaccharidic composition of applesauce is close to that of apple flesh. *Journal of Food Composition and Analysis* 24(4–5), 537–547. https://doi.org/10.1016/j.jfca.2010.12.012.

Lei, L., Yang, Y., He, H., Chen, E., Du, L., Dong, J. and Yang, J. (2016). Flavan-3-ols consumption and cancer risk: A meta-analysis of epidemiologic studies. *Oncotarget* 7(45), 73573–73592. https://doi.org/10.18632/oncotarget.12017.

Li, Q., Li, J., Duan, M., Liu, L., Fu, Y., McClements, D. J., Zhao, T., Lin, H., Shi, J. and Chen, X. (2022). Impact of food additive titanium dioxide on the polyphenol content and antioxidant activity of the apple juice. *LWT* 154, 112574. https://doi.org/10.1016/j.lwt.2021.112574.

Liu, X., Raghuvanshi, R., Ceylan, F. D. and Bolling, B. W. (2020). Quercetin and its metabolites inhibit recombinant human angiotensin-converting enzyme 2 (ACE2) activity. *Journal of Agricultural and Food Chemistry* 68(47), 13982–13989. https://doi.org/10.1021/acs.jafc.0c05064.

Lu, Y. and Yeap Foo, L. (2000). Antioxidant and radical scavenging activities of polyphenols from apple pomace. *Food Chemistry* 68(1), 81–85. https://doi.org/10.1016/S0308-8146(99)00167-3.

Lyu, F., Luiz, S. F., Azeredo, D. R. P., Cruz, A. G., Ajlouni, S. and Ranadheera, C. S. (2020). Apple pomace as a functional and healthy ingredient in food products: a review. *Processes* 8(3), 319. https://doi.org/10.3390/pr8030319.

Maldonado-Celis, M. E., Bousserouel, S., Gossé, F., Lobstein, A. and Raul, F. (2009). Apple procyanidins activate apoptotic signaling pathway in human colon adenocarcinoma cells by a lipid-raft independent mechanism. *Biochemical and Biophysical Research Communications* 388(2), 372–376. https://doi.org/10.1016/j.bbrc.2009.08.016.

Manach, C., Scalbert, A., Morand, C., Rémésy, C. and Jiménez, L. (2004). Polyphenols: food sources and bioavailability. *The American Journal of Clinical Nutrition* 79(5), 727–747. https://doi.org/10.1093/ajcn/79.5.727.

Marcillo-Parra, V., Tupuna-Yerovi, D. S., González, Z. and Ruales, J. (2021). Encapsulation of bioactive compounds from fruit and vegetable by-products for food application: a review. *Trends in Food Science and Technology* 116, 11–23. https://doi.org/10.1016/j.tifs.2021.07.009.

Marcotte, B. V., Verheyde, M., Pomerleau, S., Doyen, A. and Couillard, C. (2022). Health benefits of apple juice consumption: a review of interventional trials on humans. *Nutrients* 14(4), 821. https://doi.org/10.3390/nu14040821.

Miller, V., Yusuf, S., Chow, C. K., Dehghan, M., Corsi, D. J., Lock, K., Popkin, B., Rangarajan, S., Khatib, R., Lear, S. A., Mony, P., Kaur, M., Mohan, V., Vijayakumar, K., Gupta, R., Kruger, A., Tsolekile, L., Mohammadifard, N., Rahman, O., Rosengren, A., Avezum, A., Orlandini, A., Ismail, N., Lopez-Jaramillo, P., Yusufali, A., Karsidag, K., Iqbal, R., Chifamba, J., Oakley, S. M., Ariffin, F., Zatonska, K., Poirier, P., Wei, L., Jian, B., Hui, C., Xu, L., Xiulin, B., Teo, K. and Mente, A. (2016). Availability, affordability, and consumption of fruits and vegetables in 18 countries across income levels: findings from the Prospective Urban Rural Epidemiology (PURE) study. *The Lancet. Global Health* 4(10), e695–e703. https://doi.org/10.1016/S2214-109X(16)30186-3.

Mossine, V. V., Mawhinney, T. P. and Giovannucci, E. L. (2020). Dried fruit intake and cancer: A systematic review of observational studies. *Advances in Nutrition* 11(2), 237–250. https://doi.org/10.1093/advances/nmz085.

Musacchi, S. and Serra, S. (2018). Apple fruit quality: overview on pre-harvest factors. *Scientia Horticulturae* 234, 409–430. https://doi.org/10.1016/j.scienta.2017.12.057.

Nasri, H., Baradaran, A., Shirzad, H. and Rafieian-Kopaei, M. (2014). New concepts in nutraceuticals as alternative for pharmaceuticals. *International Journal of Preventive Medicine* 5(12), 1487–1499.

OECD (2019). *Safety Assessment of Foods and Feeds Derived from Transgenic Crops, Volume 3. Common Bean, Rice, Cowpea and Apple Compositional Considerations,*

Novel Food and Feed Safety, Organization for Economic Co-operation and Development Publishing, Paris.

Patocka, J., Bhardwaj, K., Klimova, B., Nepovimova, E., Wu, Q., Landi, M., Kuca, K., Valis, M. and Wu, W. (2020). *Malus domestica*: a review on nutritional features, chemical composition, traditional and medicinal value. *Plants* 9(11), 1408. https://doi.org/10.3390/plants9111408.

Penny, M. E., Meza, K. S., Creed-Kanashiro, H. M., Marin, R. M. and Donovan, J. (2017). Fruits and vegetables are incorporated into home cuisine in different ways that are relevant to promoting increased consumption. *Maternal and Child Nutrition* 13(3), e12356. https://doi.org/10.1111/mcn.12356.

Pollini, L., Cossignani, L., Juan, C. and Mañes, J. (2021). Extraction of phenolic compounds from fresh apple pomace by different non-conventional techniques. *Molecules* 26(14), 4272. https://doi.org/10.3390/molecules26144272.

Preti, R. and Tarola, A. M. (2021). Study of polyphenols, antioxidant capacity and minerals for the valorisation of ancient apple cultivars from Northeast Italy. *European Food Research and Technology* 247(1), 273–283. https://doi.org/10.1007/s00217-020-03624-7.

Pu, Y., He, X., Chen, L., Wang, H., Ma, Y. and Jiang, W. (2022). Apple polyphenols attenuate the binding ability of angiotensin converting enzyme 2 to viral proteins: computer simulation and in vitro experiments. *Food Bioscience* 50, 102090. https://doi.org/10.1016/j.fbio.2022.102090.

Pucci, G., Alcidi, R., Tap, L., Battista, F., Mattace-Raso, F. and Schillaci, G. (2017). Sex- and gender-related prevalence, cardiovascular risk and therapeutic approach in metabolic syndrome: a review of the literature. *Pharmacological Research* 120, 34–42. https://doi.org/10.1016/j.phrs.2017.03.008.

Rekhy, R. and McConchie, R. (2014). Promoting consumption of fruit and vegetables for better health. Have campaigns delivered on the goals? *Appetite* 79, 113–123. https://doi.org/10.1016/j.appet.2014.04.012.

Rothwell, J. A., Perez-Jimenez, J., Neveu, V., Medina-Remón, A., M'hiri, N., García-Lobato, P., Manach, C., Knox, C., Eisner, R., Wishart, D. S. and Scalbert, A. (2013). Phenol-Explorer 3.0: a major update of the phenol-Explorer database to incorporate data on the effects of food processing on polyphenol content. *Database: the Journal of Biological Databases and Curation* 2013, bat070. https://doi.org/10.1093/database/bat070.

Rudell, D. R., Buchanan, D. A., Leisso, R. S., Whitaker, B. D., Mattheis, J. P., Zhu, Y. and Varanasi, V. (2011). Ripening, storage temperature, ethylene action, and oxidative stress alter apple peel phytosterol metabolism. *Phytochemistry* 72(11–12), 1328–1340. https://doi.org/10.1016/j.phytochem.2011.04.018.

Rupasinghe, H. P. V., Thilakarathna, S. and Sandhya, N. (2012). "Polyphenols of apples and their potential health benefits". In: Sun, J., Prasad, K. N., Ismail, A., Yang, B., You, X. and Li, L. (Eds). *Polyphenols: Chemistry, Dietary Sources and Health Benefits*, Nova Science Publishers, New York, pp. 333–368. Available at: https://www.novapublishers.com/wp-content/uploads/2019/10/978-1-62081-809-1_ch16.pdf.

Salazar-Orbea, G. L., García-Villalba, R., Bernal, M. J., Hernández-Jiménez, A., Egea, J. A., Tomás-Barberán, F. A. and Sánchez-Siles, L. M. (2023). Effect of storage conditions on the stability of polyphenols of apple and strawberry purees produced at industrial scale by different processing techniques. *Journal of Agricultural and Food Chemistry* 71(5), 2541–2553, https://doi.org/10.1021/acs.jafc.2c07828.

Sandoval-Ramírez, B. A., Catalán, Ú., Calderón-Pérez, L., Companys, J., Pla-Pagà, L., Ludwig, I. A., Romero, M. P. and Solà, R. (2020). The effects and associations of whole-apple intake on diverse cardiovascular risk factors. A narrative review. *Critical Reviews in Food Science and Nutrition* 60(22), 3862–3875. https://doi.org/10.1080/10408398.2019.1709801.

Souci, S. W., Fachmann, W. and Kraut, H. (2000). *Food Compoisition and Nutrition Tables* (vol. 6), CRC Press, Stuttgart.

Šamec, D., Karalija, E., Šola, I., Vujčić Bok, V. and Salopek-Sondi, B. (2021). The role of polyphenols in abiotic stress response: the influence of Molecular Structure. *Plants* 10(1), 118. https://doi.org/10.3390/plants10010118.

Średnicka-Tober, D., Barański, M., Kazimierczak, R., Ponder, A., Kopczyńska, K. and Hallmann, E. (2020). Selected antioxidants in organic vs. conventionally grown apple fruits. *Applied Sciences* 10(9), 2997. http://doi.org/10.3390/app10092997.

USDA (2015). *National Nutrient Database for Standard Reference Release 28*. Available at: https://data.nal.usda.gov/dataset/composition-foods-raw-processed-prepared-usda-national-nutrient-database-standard-reference-release-28-0 (accessed 1.03.2023).

Vasile, M., Bunea, A., Ioan, C. R., Ioan, B. C., Socaci, S. and Viorel, M. (2021). Phytochemical content and antioxidant activity of Malus domestica Borkh Peel extracts. *Molecules* 26(24), 7636. https://doi.org/10.3390/molecules26247636.

Wallace, T. C., Bailey, R. L., Blumberg, J. B., Burton-Freeman, B., Chen, C. O., Crowe-White, K. M., Drewnowski, A., Hooshmand, S., Johnson, E., Lewis, R., Murray, R., Shapses, S. A. and Wang, D. D. (2020). Fruits, vegetables, and health: A comprehensive narrative, umbrella review of the science and recommendations for enhanced public policy to improve intake. *Critical Reviews in Food Science and Nutrition* 60(13), 2174–2211. https://doi.org/10.1080/10408398.2019.1632258.

Wang, S., Moustaid-Moussa, N., Chen, L., Mo, H., Shastri, A., Su, R., Bapat, P., Kwun, I. and Shen, C. L. (2014). Novel insights of dietary polyphenols and obesity. *The Journal of Nutritional Biochemistry* 25(1), 1–18. https://doi.org/10.1016/j.jnutbio.2013.09.001.

Wojdyło, A., Oszmiański, J. and Laskowski, P. (2008). Polyphenolic compounds and antioxidant activity of new and old apple varieties. *Journal of Agricultural and Food Chemistry* 56(15), 6520–6530. https://doi.org/10.1021/jf800510j.

World Health Organization (2003). Fruit and Vegetable Promotion Initiative: Report of the Meeting, Geneva. Available at: https://apps.who.int/iris/handle/10665/68395 (accessed 1.03.2023).

World Health Organization (2019). Increasing fruit and vegetable consumption to reduce the risk of noncommunicable diseases. Available at: https://www.who.int/elena/titles/fruit_vegetables_ncds/en/ (accessed 1.03.2023).

Yang, D., Wang, T., Long, M. and Li, P. (2020). Quercetin: its main pharmacological activity and potential application in clinical medicine. *Oxidative Medicine and Cellular Longevity* 2020, 8825387. https://doi.org/10.1155/2020/8825387.

Zhao, S., Bomser, J., Joseph, E. L. and DiSilvestro, R. A. (2013). Intakes of apples or apple polyphenols decease plasma values for oxidized low-density lipoprotein/beta2-glycoprotein I complex. *Journal of Functional Foods* 5(1), 493–497. https://doi.org/10.1016/j.jff.2012.08.010.

Zeb, A. (2020). Concept, mechanism, and applications of phenolic antioxidants in foods. *Journal of Food Biochemistry* 44(9), e13394. https://doi.org/10.1111/jfbc.13394.

Chapter 4

Advances in understanding the development of antioxidant nutraceutical compounds in apples

Matteo Scampicchio, Free University of Bolzano, Italy

1 Introduction

Apples are particularly nutrient-rich fruits used in most diets around the world. In addition to being an excellent source of nutrients, apples are also an important source of antioxidants. These compounds not only contribute to fruit organoleptic qualities but also provide functional and health benefits. This chapter explores the role of apple antioxidants, covering aspects ranging from their metabolism during the fruiting process to their stability during harvest and their subsequent bioavailability during consumption. In the first part of this chapter, we delve into the biochemical and genetic factors that influence the production of antioxidants in apple. Next, the effects of ripening and subsequent

http://dx.doi.org/10.19103/AS.2023.0127.06

harvesting of apples on changes in antioxidant properties are discussed. Another important aspect of this chapter is the health benefits of apples. Much research has linked apple consumption with a reduced risk of various diseases, from cardiovascular diseases to some types of cancer. However, the effectiveness of these benefits is closely linked to the bioavailability of antioxidants, a topic that is still poorly understood and deserves further investigation.

This chapter provides an overview of the current knowledge on apple antioxidants, from their content in the fruit, to their mechanism of action, and finally to their role in human health. Overall, we hope to emphasize the importance of apples not only as a fruit and a powerful source of nutraceutical compounds that can promote human health.

2 Nutritional composition of apples: macronutrients and micronutrients

Apples are consumed all over the world not only because of their taste but also because they are rich in essential nutrients. The nutritional profile of an apple can vary because of several factors, such as variety, environment, orchard management, and type of post-harvest technology. Apples are a great source of macronutrients, micronutrients, and phytochemicals. A complete breakdown of these nutrients is given in the following sections (see also Table 1). This section focuses on macro- and micronutrients.

2.1 Macronutrients

Macronutrients refer to the nutrients that our body needs in substantial amounts to generate energy and support overall growth. This category includes carbohydrates, dietary fiber, proteins, and fats (Agnolet et al., 2017).

2.1.1 Carbohydrates

A significant component of apples is carbohydrates. A medium-sized apple weighing approximately 180 g typically has approximately 25 g of carbohydrates, translating to approximately 95 calories. The inherent sweetness of apples comes from their soluble sugars (one group of soluble carbohydrates), usually between 10 and 15°Brix. This measure of soluble sugar denoted as °Brix) can differ among apple varieties. For example, 'Fuji' apples tend to have a higher sugar content than the 'Granny Smith' variety (Commisso et al., 2021).

2.1.2 Dietary fiber

Dietary fiber plays a pivotal role in maintaining digestive health. They foster a healthy gut, assist in managing weight, and can mitigate the risk of certain diseases such as heart disease, stroke, type 2 diabetes, and bowel cancer

Table 1 Approximate nutritional composition of apples

Nutrient	Per 100 g serving	Per medium-sized apple (180 g)	Apple juice (per 100 g)
Macronutrients			
Total lipids (g)	0.2	0.36	0.13
Saturated fat (g)	0.03	0.054	n.d.
Total carbohydrates (g)	13.8	24.84	11.3
Dietary fiber (g)	2.4	4.32	0.2
Sugars (g)	10.3	18.54	9.62
Protein (g)	0.3	0.54	0.1
Micronutrients			
Vitamin C (mg)	4.6	8.28	1
Vitamin A (IU)	54	97.2	21
Vitamin K (μg)	2.2	3.96	
Vitamin B6			0.018
Thiamin			0.021
Riboflavin			0.017
Folate (μg)	3	5.4	0
Potassium (mg)	107	192.6	101
Magnesium (mg)	5	9	5
Calcium (mg)	8	10.8	8
Phosphorus (mg)	11	19.8	7

(Su et al., 2022). An apple typically contains approximately 3 g of dietary fiber. Approximately 25% is soluble fiber, whereas the rest is insoluble (Iqbal et al., 2022). This proportion of soluble to insoluble fibers in apples makes them particularly beneficial for the digestive system.

2.1.3 Proteins and lipids

Although apples are a good source of carbohydrates and dietary fibers, they have relatively low protein and fat contents. An apple typically contains approximately 0.5 g of protein and approximately 0.3 g of fat. Although these macronutrients are present in minor amounts, they still contribute to the apple's overall nutritional value.

2.2 Micronutrients

Micronutrients are required in smaller amounts and play crucial roles in various physiological functions.

2.2.1 Vitamin C

One of the most recognized characteristics of apples is their rich source of vitamin C. Ascorbic acid, commonly known as vitamin C, is a powerful antioxidant primarily because of its electron-donating ability (Kamrul et al., 2016). This ability neutralizes unstable free radicals, thereby preventing cellular damage. In addition, ascorbic acid can work in synergy with other antioxidants, such as vitamin E, enhancing their protective effects. Its capacity to chelate metal ions further reduces the potential for free radical generation. Ascorbic acid also plays a specific role in human health by neutralizing various reactive oxygen species (ROS), alleviating oxidative stress status in apple cells and in the human body, and strengthening the immune system by aiding cellular functions (Fang et al., 2020b).

Depending on the variety, vitamin C content can range from 2 mg/100 g to 35 mg/100 g (Bassi et al., 2018). Such a high concentration is not only means a high antioxidant capacity but also supports other functions essential for our health, such as being a cofactor for enzymes that play a role in collagen synthesis, peptide hormone amidation, and tyrosine metabolism (Kuiper and Vissers, 2014). A lack of this vitamin can result in delayed wound healing and bleeding and, in extreme cases, lead to scurvy (Dresen et al., 2023).

2.2.2 Minerals

Minerals are essential inorganic nutrients that are crucial for various physiological functions. Apples are notably rich in minerals such as potassium (averaging 107 mg per 100 g), calcium (averaging 6 mg per 100 g), and magnesium (averaging 5 mg per 100 g) (Boespflug et al., 2018).

3 Phytochemicals in apples: phenolic acids

Phytochemicals are active compounds in plants. Regular consumption of these compounds offers numerous health benefits. Although they are not essential for basic human survival, they are associated with several health-promoting properties. The most important group of phytochemicals in apples is polyphenols (Table 2). On average, apples contain 150 mg catechin equivalents per 100 g of edible FW (mg/100 g FW). The following discussion reviews the main classes of polyphenols found in apples, starting with phenolic acids.

The simplest polyphenols are phenolic acids. This group of secondary metabolites is widely found in fruits and vegetables. Phenolic acids can be divided into two main subclasses:

- Hydroxybenzoic acids (C6-C1)
- Hydroxycinnamic acids (C6-C3)

Table 2 Total phenolic content (TPC, mg gallic acid equivalent/100 g) and total flavonoid content of apples (TFC, mg catechin equivalent/100 g)

Variety	TPC	TFC
Fuji	231	108
Red Delicious	205	99
Northern Spy	200	96
Fortune	197	94
Gala	191	93
Liberty	179	87
Rome Beauty	159	77
Golden Delicious	152	73
Jonagold	126	62
NY647	120	61
Idared	119	56
Cortland	115	50
Empire	111	48

Source: Boyer and Liu, 2004.

R1	R2	R3	Benzoic acids
H	H	H	Benzoic acid
OH	OH	OH	Gallic acid
OH	OH	H	Protocatechuicacid
H	OH	H	p-Hydroxybenzoicacid
OCH3	OH	OCH3	Syringic acid
OCH3	OH	H	Vanillic acid

R1	R2	R3	R4	Cinnamic acids
H	H	H	H	Cinnamic acid
OH	OH	H	H	Caffeic acid
H	OH	H	H	P-Coumaricacid
OCH3	OH	H	H	Ferulic acid
OCH3	OH	OCH3	H	Sinapic acid
OH	OH	H	Quinate	Chlorogenicacid

Figure 1 Structures of common phenolic acids: benzoic acid, hydroxybenzoic acid and hydroxycinnamic acid.

Their chemical structures are shown in Fig. 1.

The main hydroxybenzoic acids in apples include: gallic, protocatechuic, vanillic, and syringic acids, all of which originate from benzoic acid. The latter is characterized by a benzene ring attached to a prop-2-enoic acid segment (-CH=CH-COOH).

The main cinnamic acids in apples include: chlorogenic acid (also known as 5'-caffeoylquinic acid), *p*-coumaric acid, ferulic acid, and sinapic acid. Chlorogenic acid, with an average concentration of 20 mg per 100 g fresh weight, is the most important, constituting half of the total hydroxycinnamates (Vrhovsek et al., 2004). However, such values indicate that variability among cultivars is approximately 50–100% relative to the average outlined here.

Chlorogenic acid appears to be indirectly responsible for the enzymatic browning of apples because it forms a brown quinone when oxidized (Cebulj et al., 2023). This is opposite to the function of ascorbic acid, which does not produce color when oxidized to dehydroascorbic acid or further degraded. Dessert apples, such as 'Nicogreen', which have a negligible content of chlorogenic acid (1–2 mg/100 g FW) and a very high content of vitamin C, resist browning when cut. In contrast 'Red Delicious' varieties contain more than 30 mg/100 g FW of chlorogenic acid and are therefore more vulnerable to browning.

The antioxidant properties of phenolic acid mainly reflect the number and arrangement of hydroxyl and methyl groups around the aromatic ring. Ferulic acid, e.g. is a stronger radical scavenger than *p*-coumaric acid. This is because ferulic acid has an additional methoxyl group that makes it more effective. However, sinapic acid is even stronger than ferulic acid because it has a further additional hydroxyl group around the aromatic ring. Chlorogenic acid, one of the most important cinnamic acids in apples, stands out as the strongest antioxidant because it has two adjacent hydroxyl groups that can scavenge twice as many harmful radical species compared with the other monohydroxy cinnamic species. The structure–activity relationships between antioxidants have been explored in other studies (Platzer et al., 2022; Preedy and Patel, 2022; Rice-Evans et al., 1996).

4 Phytochemicals in apples: flavonoids

Flavonoids are secondary metabolites of fruits and vegetables and are abundant in apples. They are responsible for color, fragrance, and flavor characteristics. They are also receiving increasing attention for their antioxidant properties. Structurally, these flavonoids consist of two aromatic rings (A and B) interconnected by a three-carbon bridge that forms a pyran ring (ring C). This basic skeleton structure, often represented as C6–C3–C, is called 'flavan'. Variations in the pyran ring lead to distinct flavonoid classes, such as flavan-3-ols, flavones, flavonols, flavanones, isoflavones, and anthocyanins (Shen et al., 2022). Each class is also populated by the presence of hydroxyl and methyl groups around the aromatic rings, A and B.

In apples, the most important flavonoids correspond to four classes (Zhang et al., 2020b):

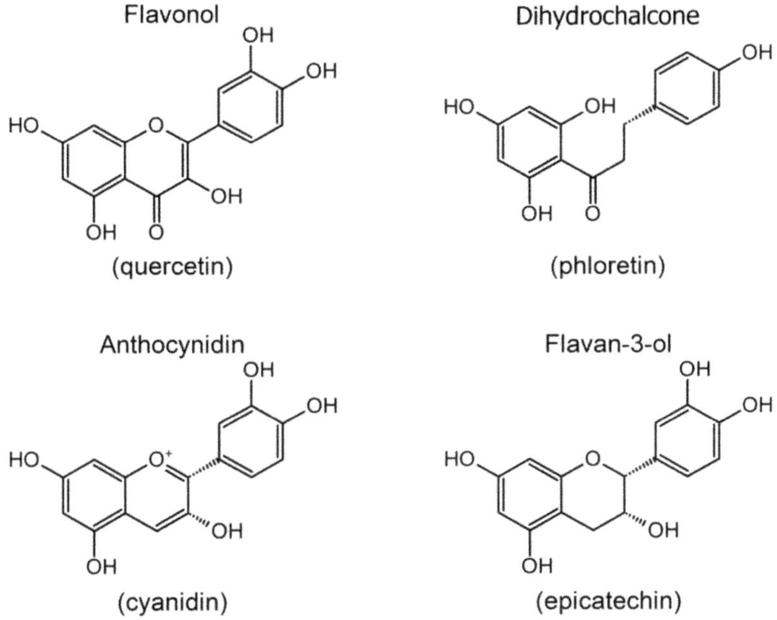

Figure 2 Structures of important flavonoids in apples: flavan-3-ol, flavonol, anthocyanidin and dihydrochalcone.

- Flavan-3-ol;
- Flavonol;
- Anthocyanidin; and
- Dihydrochalcone.

The chemical structure is shown in Fig. 2.

4.1 *Flavan-3-ols*

The flavan-3-ol subclass is characterized by a classical flavan-base structure, typical of any flavonoid, with the presence of a hydroxyl group on the pyran ring (C-ring), which provides its antioxidant properties. Representative examples of this class in apples include catechin, epicatechin, and dimeric procyanidins, the latter of which originate from condensation of catechins and epicatechins. Taken together, their average content is 12 mg/100 g FW, although their variation within cultivars is significant, within 1–40 mg/100 g FW (Ceymann et al., 2012). These compounds are mainly present in the peel and pulp. A particular characteristic of flavan-3-ols is their contribution to enzymatic browning. Flavan-3-ols are oxidized by polyphenoloxidase into brown quinones, which irreversibly form pigments responsible for apple browning.

4.2 Flavonols

Flavones form basic structures within the flavonoid family. Flavonols, a related group, are recognized for their antioxidant properties because of the presence of multiple hydroxyl groups. In apples, the dominant flavan-3-ols are the glycoside forms of quercetins. Quercetin, a prominent flavonol, is found in apples in concentrations ranging from 1 mg/100 g to 5 mg/100 g FW (Ceymann et al., 2012).

4.3 Anthocyanidins

Anthocyanidins, the aglycones (sugar-free counterparts) of anthocyanins, are a group of water-soluble flavonoids widely present in fruits and vegetables and are mainly responsible for their skin and flesh color, as seen in black berries and red-flesh apples. This class of flavonoid has resonance structures in the pyran ring (C ring), which is vital for the coloration of many plants and fruits. The principal anthocyanidins in apples are glycosylated cyanidins found at low concentrations (Wu and Prior, 2005).

4.4 Dihydrochalcones

Dihydrochalcones share a C6–C3–C6 structure with flavonoids. However, unlike flavonoids, their C3 chain remains open, without forming a pyran ring. Although these species are rarely found in most fruits, they are characteristic molecules in apples. Examples include phloretin and its glucoside derivative, phloridzin. Taken together, their content is approximately 6 mg/100 g FW, with considerable exceptions to antient varieties (Ceymann et al., 2012). There is evidence that dihydrochalcones have a significant ability to show radical scavenging activity, for instance, in suppressing lipid peroxidation (Nakamura et al., 2003).

5 Phytochemicals in apples: antioxidants in essential oils

Essential oils (EOs) are aromatic liquid mixtures of volatile compounds used for their health benefits and recognized for their antiseptic, antioxidant, and anti-inflammatory properties. Apple seed oils are mainly extracted from seeds, which contain approximately 20% oil. This oil is rich in fatty acids such as linoleic (omega-6) and oleic (omega-9) acids (Ferrentino et al., 2020b), which represents a valuable nutraceutical by-product. In addition, EOs contain minor bioactive components, such as phytosterols (β-sitosterol) and tocopherols, which contribute to health benefits. Phytosterols have been associated with cholesterol reduction, thus aiding heart health (Gupta et al., 2011), whereas tocopherols, known for their antioxidant properties, protect against lipid peroxidation and have potential anticancer effects (Jiang, 2017).

However, EOs extracted from apple seeds may also contain amygdalin, which can produce cyanide when metabolized. Although the quantity of amygdalin is harmless, especially considering the small consumption of the oils, recent studies have found that extraction with supercritical fluids can lead to apple seed oils free from this antinutrient (Ferrentino et al., 2020a).

In addition to tocopherols, the antioxidant capacity of EOs can be attributed to terpenes and phenolic compounds (Amorati et al., 2013). Despite some limitations in the extraction process, which may reduce the phenolic content, the high content of unsaturated fatty acids, phytosterols, and tocopherols, along with its antioxidant properties, presents significant health and technological applications in food preservation and disease prevention (Asma et al., 2023).

6 Distribution of antioxidants in apples

The nutritional composition of apples can vary according to the parts of the apple. For instance, the peel often has a higher antioxidant content than the flesh. A detailed examination of apple anatomy shows that apple peel is a potent source of antioxidants, often having greater concentrations of these beneficial compounds than the flesh (Kumari et al., 2023). Figure 3 shows the composition of antioxidants in different parts of apples.

Peel is the primary store of phenolic compounds in apples. In some cases, the concentration of these compounds in the peel can be nearly ten times higher than that in the flesh. The phenolic profile of apple peel varies with variety. Peel is typically rich in flavan-3-ols and flavonols. The vibrant red hue of some apple skins is attributed to anthocyanidins, with red varieties producing these compounds in higher quantities than their green or yellow counterparts.

Flesh + Peel	mg/g DW
(+)-catechin	0.4-1.5
()-Epicatechin	0.4-1.3
Rutin	0.4-1.6
Phloridzin	0.1-0.4
Chlorogenic acid	0.2-2.3
Caffeic acid	1.0-3.5

Peel	mg/g DW
(+)-catechin	0.1-0.4
()-Epicatechin	0.1-0.5
Rutin	0.25-0.9
Phloridzin	0.06-0.2
Chlorogenic acid	0.02-0.2

Seeds	mg/g DW
(+)-catechin	0.4-1.5
()-Epicatechin	0.4-1.3
Rutin	0.4-1.6
Phloridzin	0.1-0.4
Chlorogenic acid	0.2-2.3

Figure 3 Ranges of polyphenol concentrations in seed, peel and peel + flesh (Duda-Chodak et al., 2011; Francini and Sebastiani, 2013; Łata et al., 2009).

While most phenolic compounds are more abundant in the peel, chlorogenic acid is more prevalent in the flesh of specific apple types such as 'Fuji' and 'Golden Delicious'. The flesh predominantly contains compounds such as flavan-3-ols, flavonols, and dihydrochalcones. Some newer red-fleshed apple breeds have high flavonoid content (Fang et al., 2020a).

Although often overlooked, apple seeds are also rich in phenolic compounds such as chlorogenic acid and phloridzin. Phloridzin constitutes up to 90% of the total phenolic content of seeds. This unique composition underscores the potential value of apple seeds.

Traditional apple breeds possess phenolic content that is several times higher than that of popular commercial varieties such as 'Golden Delicious' and 'Red Delicious'. Some varieties such as 'Fialka' and 'Discovery' are particularly rich in phenolic content. While many varieties have minimal anthocyanin content, varieties with red peels, such as Red Delicious, are exceptions (Fang et al., 2020a).

7 Oxidative damage in apples and its impact on quality

Oxidative damage in apples is the damage caused to cells, tissues, or molecules in fruit by ROS and other free radicals (Meitha et al., 2020; Pisoschi et al., 2021). These oxygen derivative reactive molecules are formed from oxygen to create singlet oxygen. This is a highly reactive electrophilic species, which leads to various reactive forms, such as superoxide (O_2^-), hydrogen peroxide (H_2O_2), and the highly reactive hydroxyl radical (-HO•). These species can attack various cellular components, including lipids, proteins, carbohydrates, and nucleic acids (Halliwell et al., 2021).

One of the effects of oxidative damage in apples is browning, which occurs when the fruit is cut or bruised (Bodner and Scampicchio, 2020). This is a result of the oxidation of phenolic compounds, leading to the formation of brown pigments. While this does not necessarily make the apple harmful to eat, it does affect its visual appeal (Arnold and Gramza-Michałowska, 2022).

Oxidative damage can alter the cellular structure of apples, affecting their crispness and overall texture (Huang et al., 2023). The breakdown of cellular membranes can release cell contents, triggering oxidative stress and further degrading the texture of the fruit. ROS, which are responsible for damaging cell membrane components and consequently affecting the crispness of the fruit, can be generated during any phase of fruit development. In addition, enzymatic browning, lipid peroxidation, disruptions in metabolic processes, and the inability of damaged cells to regulate water content contribute to altered texture. As a result, apples subjected to oxidative stress may lose their characteristic crispness and become drier and mealy (Pieczywek et al., 2023).

Proper storage and handling can mitigate these effects and preserve the apple texture.

The oxidation of certain compounds in apples can lead to the development of off-flavors. The oxidation of α-farnesene in apples can result in superficial scald, manifesting as brown or black patches on the skin of the fruit. This indicates that an apple subjected to oxidative stress, leading to the accumulation of specific oxidation products, might not only have a compromised visual appeal but could also potentially exhibit altered flavors, making it less palatable than an apple that has been adequately stored and preserved from oxidation (Espino-Díaz et al., 2016; Whitaker, 2004).

Similar to other foods, the oxidative processes in apples can degrade essential nutrients. The antioxidant vitamin C content in apples can be reduced by oxidation, thus decreasing the nutritional value of the fruits. Oxidative damage can compromise the sensory, nutritional, and textural qualities of apples. This underscores the importance of proper storage and handling of apples to minimize exposure to factors that can accelerate oxidative damage.

8 Preventing oxidative damage: antioxidant mechanisms and measurement

Antioxidants in apples play a crucial role in inhibiting lipid oxidation, a process that is harmful to both food quality and health. This process of inhibition involves a complex series of reactions that can be simplified into three main steps:

- Initiation;
- Propagation; and
- Termination.

Figure 4 shows this process.

During the initiation phase, reactive species (often radicals) interact with oxidizable substrates, such as lipids, to form radicals. In the propagation step, these radicals react with oxygen to form peroxyl radicals, further propagating the chain reaction and producing additional radicals. This cycle continues, leading to increased oxidation and subsequent oxidative damage. Figure 5 shows the crucial role of antioxidants in preventing oxidation.

Antioxidants can inhibit the formation of initial radical species (preventive antioxidants) or react with peroxyl radicals more rapidly than the oxidizable substrate, thus breaking the chain reaction (chain-breaking antioxidants). Some bioactives, such as terpenes, are also believed to break the termination step (Amorati et al., 2013). Phenols, commonly found in apples, are chain-breaking antioxidants. They donate hydrogen atoms to peroxyl radicals, converting them into non-radical species and thereby terminating the oxidation chain.

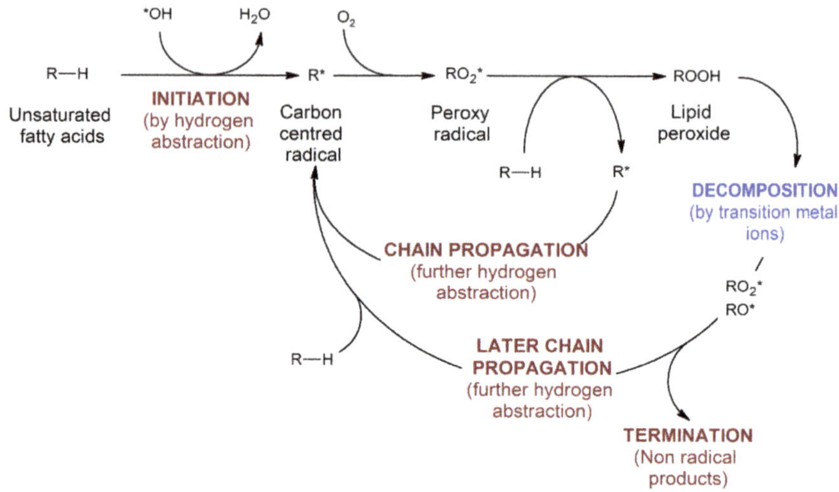

Figure 4 Autoxidation of an organic substrate and the mechanism of interference by direct antioxidants.

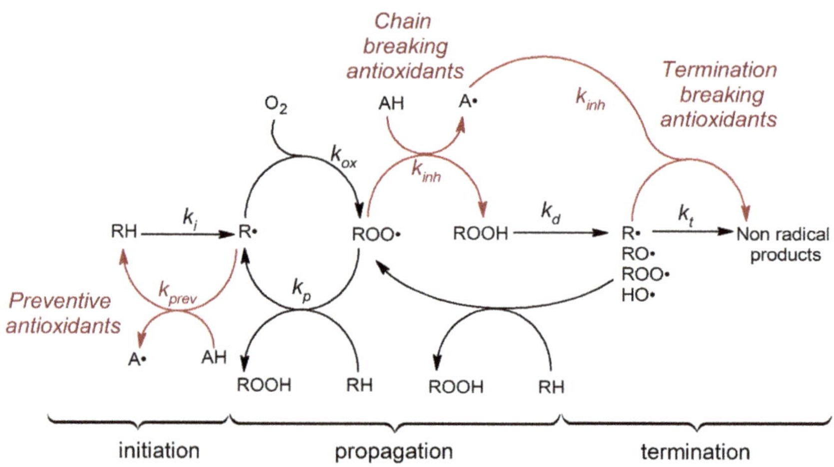

Figure 5 Role of antioxidants in preventing oxidation.

The measurement of antioxidant properties in apples uses several methods, each designed to assess different aspects of antioxidant activity. Spectrophotometric assays such as DPPH, ORAC, FRAP, and ABTS assess the capacity of antioxidants to scavenge free radicals or donate electrons (Amorati and Valgimigli, 2015). Oximetry methods measure the oxygen consumption rate in a system to determine the efficacy of chain-breaking antioxidants (Amorati and Valgimigli, 2018). In addition, methods using nanoparticles offer innovative ways to evaluate antioxidant properties, exploiting the unique interactions between antioxidants and nanoparticle surfaces (Scampicchio et al., 2006). Recently, spectrophotometric stopped flow assays have provided a fast and accurate alternative to classical assays (Angeli et al., 2021). In addition, calorimetry-based assays offer a new, robust, and high-throughput assessment of antioxidants (Mosibo et al., 2022; Suhag et al., 2024).

Despite the availability of various methods to study antioxidant properties, one of the major difficulties in the objective evaluation of antioxidants is the lack of standardized procedures (Angeli et al., 2023). The absence of standard protocols leads to inconsistent and non-comparable results among different laboratories. It should also be noted that most assays used today are limited to the measurement of antioxidant capacity alone, neglecting the actual reactivity of antioxidants with radicals. This approach can be misleading in that it does not accurately reflect the effectiveness of antioxidants in inhibiting the process of lipid oxidation. The distinction between antioxidant capacity and activity is critical (Table 3). Note that the rate at which antioxidants react with the radical species expresses antioxidant activity. Measuring activity provides a more realistic assessment than antioxidant capacity and allows better prediction of antioxidant performance in biological systems or food matrices (Apak, 2019).

Understanding the mechanisms by which antioxidants inhibit lipid oxidation in apples, along with accurate measurement of their efficacy, is essential to advance the development of nutraceutical compounds. By overcoming current challenges in standardizing measurements and focusing on antioxidant activity, we can provide a more accurate picture of the role of antioxidants in preserving food quality and promoting health. This understanding is critical for harnessing the full potential of antioxidants in apples for nutraceutical applications.

9 Pre-harvest changes in antioxidant content: maturation and ripening

A series of complex biochemical changes occur during apple maturation. As apples move from their flowering stage to full maturity, there is a notable increase in their sugar content, which includes soluble solids. Conversely, their acidity, as measured by titratable acidity and pH, decreases. Specifically, for 'Fuji' apples, there is a marked decline in antioxidant activities and polyphenol content around the 85th day after full bloom (DAFB). This decline is closely associated with the chlorogenic acid content. By the 60th DAFB, antioxidant activities reduce by half, and this decline continues past the 85th DAFB, stabilizing only after the 120th DAFB (Zheng et al., 2012)

Following the maturation process, apples undergo ripening, which results in significant physiological changes. During this phase, starch in apples is converted into sugars, and the tissues soften due to the degradation of cell walls. This softening process, combined with the development of volatile

Table 3 Traditional methods for assessment of antioxidant capacity

Class	Methods	Working principle	Conditions	Detection
HAT	ORAC assay	Reaction of fluorescein with radical initiator AAPH.	Phosphate buffer, pH = 7.2, T = 37°C	Fluorescence (excitation at 485 nm/emission at 525 nm)
HAT	TRAP assay	Reaction of R-phycoerythrin (R-PE) with AAPH	Phosphate buffer, pH = 7.0, T = 37°C	Fluorescence (excitation at 495 nm/emission at 575 nm)
HAT	Total radical scavenging capacity (TOSC) assay	Reaction between oxy-radicals and a-keto-g-methiol-butyric acid (KMBA) with production of ethylene	Potassium phosphate buffer, pH = 7.4, T = 34°C/39°C	Gas chromatography
HAT	Crocin bleaching (CB) assay	Competitive reaction between crocin, and antioxidant compounds for hunting peroxyl radicals generated by azo-initiator	Potassium phosphate buffer, pH = 7.01, T = 40°C	Visible spectrometry (443 nm)
HAT	Chemiluminescence assay	Reaction between luminol and an oxidant (H_2O_2) with/without a metal or enzymatic catalyst	Potassium phosphate buffer, pH = 7.4, T = 25°C	Chemiluminescence (689 nm)
ET	FRAP assay	Reduction of the ferric 2,4,6-tripyridyl-s-triazine complex (Fe^{3+}–TPTZ) to the blue-colored ferrous ions (Fe^{2+}–TPTZ) by antioxidants in acidic medium.	Acetate buffer, pH = 3.6, T = 37°C	Visible spectrometry (593 nm)

ET	Total phenol Folin–Ciocalteu (TFC) assay	Reduction of FC reagent by antioxidants	Sodium carbonate solution, pH = ~10, T = 23°C	Visible spectrometry (760 nm)
ET	TEAC or ABTS assay	Scavenging of ABTS radical ($ABTS^{\cdot}$), which converts into a colorless product	Ethanol/phosphate buffer, pH = 7.4, T = 30°C	Visible spectrometry (734 nm)
ET	DPPH radical scavenging assay	Delocalization of violet DPPH radical solution in presence of antioxidants	Ethanol/methanol, pH = 5.5, T = 37°C	Visible spectrometry (515 nm)
ET	*N*,*N*-Dimethyl-*p*-phenylenediamine radical scavenging (DMPD) assay	Conversion of DMPD to the colored DMPD radical cation ($DMPD^{\cdot +}$) in the presence of ferric ions (Fe^{3+})	Acetate buffer, pH = 7.4, T = 37°C	Visible spectrometry (517.4 nm)
ET	CUPRAC assay	Reduction of Cu^{2+}-neocuproine reagent to Cu^{+} complex.	Ammonium acetate buffer, pH = 7.0, T = 37°C	Visible spectrometry (450 nm)

Source: Haque et al., 2021.

compounds, contributes to the characteristic flavor of apples. In addition, apple color transforms because of the degradation of chlorophyll and the synthesis of anthocyanins. There is a general decrease in hydroxycinnamic acids, leading to the formation of various flavonoids. Among these flavonoids, dihydrochalcones are significant, with phloridzin being the most abundant dihydrochalcone, predominantly found in the leaves, bark, and fruit itself (Treutter, 2001).

Thus, the synthesis and regulation of phenolic compounds in apples are influenced by a combination of genetic, environmental, and developmental factors. As has been noted, some traditional apple varieties contain higher concentrations of beneficial phenolics than certain commercial varieties (Preti and Tarola, 2021). Environmental conditions, such as light intensity, temperature, and ultraviolet radiation, play a crucial role in affecting antioxidant content. Apples exposed to UV light tend to increase their flavonoid production as a protective measure. The geographical location where apples are grown also affects their antioxidant content, which depends on variations in temperature and the duration of exposure to sunlight (Geleta and Heo, 2023). Biotic stresses, such as those from pathogens or physical injuries, can lead to an increased production of phenolic compounds, enhancing the plant's defense mechanisms (de la Rosa et al., 2019). The method of production can also influence the antioxidant content. For example, apples grown organically have shown approximately 25% higher antiradical activity in their peel than those grown conventionally. Apple size can also be a factor, with larger apples potentially having diluted antioxidant concentrations (Vrhovsek et al., 2004).

10 Post-harvest changes in the antioxidant content

The post-harvest stage refers to the various processes and treatments that apples undergo after being harvested. This stage encompasses storage, handling, and post-harvest treatments to preserve freshness, prevent spoilage, and enhance certain properties of the fruit (Dixon and Hewett, 2010). The post-harvest stage is a critical phase in the life cycle of apples, during which the fruit undergoes various biochemical transformations that ensure that the taste, texture, and nutritional benefits of apples are retained during transportation and storage. The post-harvest stage can also significantly affect the antioxidant content of apples.

One of the primary post-harvest reactions affecting antioxidants is oxidation (Meitha et al., 2020). ROS, which are generated during post-harvest handling and storage, induce oxidative stress, leading to the degradation of antioxidants (Chen et al., 2023). Enzymatic activities, especially those catalyzed by enzymes such as polyphenol oxidase and peroxidase, also play a pivotal role in the breakdown of antioxidant compounds, further diminishing their concentration in the fruit (Arnold and Gramza-Michałowska, 2022; Serra et al., 2021).

External factors, including exposure to light, heat, and oxygen, can also exacerbate these changes. Light exposure can induce photochemical reactions that degrade specific antioxidants such as ascorbic acid (Yin et al., 2022). Elevated temperatures can also stimulate enzymatic activity, accelerating the oxidation process (Arora et al., 2018). Prolonged exposure to oxygen can trigger oxidative reactions that further deplete antioxidant levels (Meitha et al., 2020).

11 Post-harvest techniques to preserve antioxidant content

Post-harvest preservation of antioxidants in apples is critical for maintaining their nutritional and health-promoting properties (Van Der Sluis et al., 2005). This section discusses a range of techniques:

- Controlled atmosphere storage;
- SmartFresh technology;
- UV radiation; and
- Edible coatings.

The method of storage employed can have a profound influence on antioxidant preservation. One of the most important storage techniques is controlled atmosphere (CA) storage. Within a regulated environment, oxygen levels are decreased and carbon dioxide levels are elevated, effectively decelerating the fruit respiration (Butkeviciute et al., 2022). The temperature is also maintained at a lower range, further suppressing metabolic activity (Prange and Wright, 2023). Humidity is also precisely regulated to prevent moisture loss, thereby ensuring that apples retain their textural integrity (Lee et al., 2019). Research has shown that 'Red Delicious' apples stored under CA conditions retain a significant portion of their original antioxidant activity even after 6 months, a feat not achieved under regular cold storage conditions (Shen et al., 2021).

Emerging postharvest strategies are continually being explored to optimize antioxidant preservation in apples. One such technique is the SmartFresh technology (Kolniak-Ostek et al., 2014; Prange and Wright, 2023). SmartFresh technology, scientifically known as 1-methylcyclopropene (1-MCP) treatment, is a technique that has attracted significant attention, particularly for apples (Lee et al., 2012; Watkins, 2006; Watkins et al., 2000). This technology operates by inhibiting the action of ethylene, a naturally occurring plant hormone pivotal to the ripening and senescence processes of fruits. By introducing apples to an environment enriched with 1-MCP, the binding of ethylene to its receptors is competitively obstructed, thereby delaying the ripening and aging processes.

SmartFresh thus extends the post-harvest shelf life of apples, ensuring that they retain their freshness, firmness, and overall quality for extended periods (Lee et al., 2012; Watkins et al., 1997). This technology, underpinned by rigorous scientific research, offers a sustainable and efficient solution to the challenges of postharvest fruit preservation in the global agricultural sector (Chris B. Watkins, 2006). As an example, research indicates that 'Granny Smith' apples subjected to this technology retained a significant portion of their antioxidant capacity even after 3 months of storage (Deyman et al., 2014).

It is well-known that the position of apples on the tree affects their polyphenolic content due to differential exposure to sunlight (Awad et al., 2001). This has prompted approaches aimed at regulating the formation of flavonoids and chlorogenic acids based on artificial irradiation. Although still expensive, irradiation treatments based on UV light are effective in increasing the phenolic content of apple, increasing red skin color, antioxidant capacity, and the content of various phenolic compounds, including anthocyanins and quercetin glycosides (Arnold and Gramza-Michałowska, 2022). However, UV irradiation treatments affect only the apple peel and not the flesh. This is a limitation that should be considered in the evaluation of fruit quality using this technology.

Edible coatings are thin layers of natural biopolymers applied to apples to prolong shelf life and give them a glossy finish. They can be considered an alternative technology to preserve apples during post-harvesting and an ideal replacement for paraffin treatments, which are currently prohibited in many countries (Maringgal et al., 2020). Edible coatings are safe because they are composed mainly of natural biopolymers, such as polysaccharides (Tahir et al., 2019), proteins (Perez-Gago et al., 2006), and lipids (Conforti and Totty, 2007). Their effectiveness in delaying deterioration is attributed to their barrier-effect against oxygen and moisture diffusion, which slow the ripening process (Aayush et al., 2022). Particularly beneficial in organic farming, edible coatings can also serve as carriers of flavorings or nutrients.

Several examples of edible coatings have effectively maintained the quality of apples. Examples include edible coatings made of zein and nisin (Belay et al., 2023), candelilla wax, and fermented extract of tarbush (De León-Zapata et al., 2015), chitosan coating combined with banana peel extract (Zhang et al., 2020a), or a combination of carnauba wax and 1-methylcyclopropene (1-MCP) (Chen et al., 2020). Edible coatings are effective in preserving vitamins, e.g. edible sucrose monoester coatings that could extend the stability of phenols and vitamin C (Jung and Choi, 2021).

12 Antioxidant bioavailability

The ability of our bodies to absorb and utilize antioxidants, termed bioavailability, has become a focal point of recent research (Williamson, 2017).

Before being absorbed, phenolic compounds undergo a complex process to induce biological activity after consumption. Several factors determine this bioavailability. If we define the bioavailability of nutrients as the ratio of substances introduced into the diet that are absorbed and available for use or storage, the bioavailability of phenolic compounds is limited by factors such as their release from the food matrix and the extent of absorption in the gastrointestinal tract (Velderrain-Rodríguez et al., 2014). The bioavailability of phenolic compounds also depends on the food matrix itself, including its composition, processing methods, and the presence of other compounds that might affect absorption. Finally, individual differences, such as health status, enzyme activity, and gender, can influence antioxidant absorption and use. In the case of apples, environmental factors, such as light exposure, temperature, and water stress, can also affect antioxidant bioavailability(Łata et al., 2022).

The first step (1) is the release of phenolic compounds from the apple matrix (Liu et al., 2019). Release starts with mastication and oral digestion with saliva enzymes, but it becomes effective only in the gastric phase (2), where approximately 65% of the phenolic acids and flavonoids in apples are released (Eran Nagar et al., 2020). The remaining phenolic compounds are then released into the small intestine, where they are absorbed by epithelial cells (3). Once transported into the bloodstream, phenolics can be redistributed in tissues by entering the systemic circulation. After enteral absorption, the majority of polyphenols are extensively transformed by the detoxification system in enterocytes and liver, before (4), being finally eliminated via excretion in bile, feces, and urine (Feng et al., 2021). The uptake routes for apple flavonoids during digestion are still a subject of debate, with various enzymes and transporters playing potential roles in their absorption. However, recent findings indicate that digestion is the key stage that determines polyphenol bioaccessibility (Gonçalves et al., 2019).

13 Health benefits of antioxidants in apples

This section briefly reviews the health benefits of apple antioxidants, particularly the mechanisms underlying these benefits. In general, the consumption of fruits has been linked to a reduced risk of several major diseases. This has been translated into a recommendation from the World Health Organization and the Food and Agriculture Organization to consume approximately 400 g of fruits and vegetables daily. A 15-year study involving 1456 women over 70 years found that those consuming an apple daily had a 35% lower risk of death from cancer (Hodgson et al., 2016).

Apples contain polyphenols that may scavenge free radicals (Kumari et al., 2023; Millán-Laleona et al., 2023; Williamson, 2017). Polyphenols exhibit inhibitory efficacy against 2,2-dipheynl-1-picrylhydrazylradicals

(DPPH), antioxidant activity in preventing lipid peroxidation, and the ability to scavenge ROS, including hydrogen peroxide (Kamrul et al., 2016). Such *in vitro* evidence has also been confirmed across multiple *in vivo* studies, where apple polyphenols have effectively inhibited oxidative stress markers associated with cancer and liver damage (Preedy and Patel, 2022).

13.1 Cardiovascular disease

Early research indicated a negative correlation between apple consumption, which is a source of antioxidants, and coronary mortality (Boyer and Liu, 2004). More recent epidemiological studies have confirmed that apple consumption correlates with a decline in cardiovascular disease (Hansen et al., 2010; Larsson et al., 2013). One study, e.g., found that individuals with the highest fruit and vegetable intake had a 20% reduced risk of coronary heart disease, with the most significant benefits observed in those consuming vitamin C-rich fruits (Joshipura et al., 2001). A recent overview of the role of plant phytochemicals in preventing cardiovascular disease was provided by Sadgrove and Simmonds (2022). In the case of apples, this protective effect is probably due to the ability of apple polyphenols to inhibit lipid peroxidation (Serra et al., 2012) and oxidative stress (Hyson, 2011). However, it remains uncertain whether this benefit arises from a single compound or a combined effect of all apple components.

13.2 Cancer

There is a consensus that diet could play a role in cancer prevention. It has been estimated that one-third of all cancer deaths could be prevented through correct dietary habits, notably through increased consumption of fruits, vegetables, and whole grains (Hyson, 2011).

The antioxidant activity of apples involves the capacity of bioactives in apples to neutralize harmful ROS, which can damage cellular components such as DNA (Boyer and Liu, 2004). This damage can lead to mutations that initiate the transformation of normal cells into cancerous cells. Antioxidants not only neutralize these harmful molecules but also aid in DNA repair, inhibit cancer cell growth, reduce inflammation, regulate hormones, and support the immune system (Didier et al., 2023). Apple polyphenols such as quercetin have been linked to improved endothelial function and reduced blood pressure. A study with over 6000 participants from Italy associated daily consumption of antioxidant-rich apples with decreased cancer risk across various types, including oral cavity, esophagus, and breast cancers (Gallus et al., 2005).

However, the relationship between antioxidants and cancer is complex. While a diet rich in foods containing antioxidants is often linked to reduced cancer risk, the efficacy of antioxidant supplements in cancer prevention remains uncertain. Some reports even suggest potential harm from high doses of certain antioxidant supplements (Li et al., 2022).

13.3 Aging and cognitive processes

Emerging evidence indicates that dietary choices can influence cognitive decline during aging and may also impact the risk and progression of neurodegenerative diseases, such as Alzheimer disease (Hyson, 2011). Apple juice concentrates led to improved cognitive performance, especially when subjects were exposed to high oxidative stress conditions. Such activity was attributed to molecules such as quercetin and catechin, which prevent the decline in acetylcholine, a neurotransmitter crucial for memory and cognitive performance.

13.4 Diabetes

With the alarming rise of diabetes worldwide, especially type 2 diabetes, research has intensified to find dietary solutions. Regular apple consumption, particularly 2–6 apples a week or 1 apple a day, has been associated with a 27% and 28% lower risk of type 2 diabetes, respectively (Song et al., 2005). Specific antioxidant components in apples, such as catechins, may play a crucial role in reducing diabetes risk by preserving pancreatic b-cell function.

13.5 Bone health

Bone mass loss leading to osteoporosis is a significant global health concern. Fruits and vegetables, especially apples, are rich in antioxidants and essential nutrients that may bolster bone health. Some studies have indicated that antioxidants in apples can positively impact markers related to bone health (Bell and Whiting, 2004).

13.6 Gastrointestinal protection from drug injury

Certain studies have investigated the potential of apples, which are rich in antioxidants, to shield the gastric mucosa from drug-induced injuries (Graziani et al., 2005). These studies have shown that apple extracts, especially those high in antioxidant molecules such as chlorogenic acid and catechin, can protect cells from oxidative damage and prevent drug-induced injuries.

14 Conclusion and future trends

This chapter provides an overview of the current knowledge on the development, stability, and bioavailability of apple-derived antioxidants. It has also emphasized the importance of apple antioxidants in promoting human health.

Although considerable progress has been made in understanding the antioxidant properties of apples, there are areas where further research could provide valuable insights. These include a better understanding of the genetic mechanisms that regulate antioxidant synthesis and accumulation, new genetic approaches required to design cultivars with high levels of antioxidants, further exploration of the impact of environmental and cultural conditions on antioxidant concentrations (including optimizing cultivation and post-harvest handling techniques), and the study of methods to improve the bioavailability and uptake of apple-derived antioxidants.

A growing area of interest is the use of apple processing as a source of antioxidants, such as flavonoids and phenolic acids, in the development of various food products, such as beverages, baked goods, and dairy products. A concrete example is the use of apple pomace, a by-product of apple juice production that is rich in phenolic compounds. This is used as a functional ingredient in the production of bread. Bread made with 5% apple pomace has 2.5 times more polyphenols, 8 times more flavonoids, 4 times more chlorogenic acid, and 21 times more phloridzin than the control, resulting in 6.5 times higher antioxidant potential, while still maintaining optimal organoleptic perception (Gumul et al., 2021).

The expansion of the functional food market opens new prospects for the use of apple-derived antioxidants in a broader range of food products. This expansion requires studies on the stability of these antioxidants during food processing and their influence on the sensory properties of the final product.

New methods to characterize antioxidant activity, especially on a kinetic basis, need to be developed. Indeed, without these methods, the ability to understand the genetic mechanisms that regulate the synthesis and accumulation of these compounds would be lost. Equally important are the bioavailability and absorption of apple-derived antioxidants in the human body. Extensive studies, including human clinical studies, are required to better understand these processes and improve the absorption and utilization of these antioxidants. Further clinical studies are needed to establish concrete dietary recommendations and to understand the role of apple-derived antioxidants in disease prevention and management.

15 References

Aayush, K., McClements, D. J., Sharma, S., Sharma, R., Singh, G. P., Sharma, K. and Oberoi, K. (2022). Innovations in the development and application of edible coatings for fresh and minimally processed apple. *Food Control* 141, 109188. https://doi.org/10.1016/J.FOODCONT.2022.109188.

Agnolet, S., Ciesa, F., Soini, E., Cassar, A., Matteazzi, A., Guerra, W., Robatscher, P., Storti, A., Baric, S., Dalla Via, J. and Oberhuber, M. (2017). Makroelemente und Qualitätsparameter von 34 alten und acht kommerziellen Apfelsorten angebaut am selben Standort in Südtirol, Italien. *Erwerbs-Obstbau* 59(3), 171–183. https://doi.org/10.1007/S10341-017-0336-8/FIGURES/5.

Amorati, R., Foti, M. C. and Valgimigli, L. (2013). Antioxidant activity of essential oils. *Journal of Agricultural and Food Chemistry* 61(46), 10835–10847. https://doi.org/10.1021/jf403496k.

Amorati, R. and Valgimigli, L. (2015). Advantages and limitations of common testing methods for antioxidants. *Free Radical Research* 49(5), 633–649. https://doi.org/10.3109/10715762.2014.996146.

Amorati, R. and Valgimigli, L. (2018). Antioxidant activity of essential oils. *Journal of agricultural and food chemistry* 66(13), 3324-3329. https://doi.org/10.1021/jf403496k.

Angeli, L., Imperiale, S., Ding, Y., Scampicchio, M. and Morozova, K. (2021). A novel Stoichio-kinetic model for the DPPH• assay: the importance of the side reaction and application to complex mixtures. *Antioxidants* 10(7). https://doi.org/10.3390/antiox10071019.

Angeli, L., Morozova, K. and Scampicchio, M. (2023). A kinetic-based stopped-flow DPPH• method. *Scientific Reports* 13(1). https://doi.org/10.1038/S41598-023-34382-7.

Apak, R. (2019). Current issues in antioxidant measurement. *Journal of Agricultural and Food Chemistry* 67(33), 9187–9202. https://doi.org/10.1021/acs.jafc.9b03657.

Arnold, M. and Gramza-Michałowska, A. (2022). Enzymatic browning in apple products and its inhibition treatments: a comprehensive review. *Comprehensive Reviews in Food Science and Food Safety* 21(6), 5038–5076. https://doi.org/10.1111/1541-4337.13059.

Arora, B., Sethi, S., Joshi, A., Sagar, V. R. and Sharma, R. R. (2018). Antioxidant degradation kinetics in apples. *Journal of Food Science and Technology* 55(4), 1306. https://doi.org/10.1007/S13197-018-3041-1.

Asma, U., Morozova, K., Ferrentino, G. and Scampicchio, M. (2023). Apples and apple by-products: antioxidant properties and food applications. *Antioxidants (Basel, Switzerland)* 12(7). https://doi.org/10.3390/antiox12071456.

Awad, M. A., Wagenmakers, P. S. and De Jager, A. (2001). Effects of light on flavonoid and chlorogenic acid levels in the skin of "Jonagold" apples. *Scientia Horticulturae* 88(4), 289–298. https://doi.org/10.1016/S0304-4238(00)00215-6.

Bassi, M., Lubes, G., Bianchi, F., Agnolet, S., Ciesa, F., Brunner, K., Guerra, W., Robatscher, P. and Oberhuber, M. (2018). Ascorbic acid content in apple pulp, peel, and monovarietal cloudy juices of 64 different cultivars. *International Journal of Food Properties* 20 (sup3), S2626–S2634. https://doi.org/10.1080/10942912.2017.1381705.

Belay, Z. A., Mashele, T. G., Botes, W. J. and Caleb, O. J. (2023). Effects of zein-nisin edible coating on physicochemical and microbial load of 'Granny Smith' apple after long term storage. *http://Mc.Manuscriptcentral.Com/Tcyt* 21(1), 334–343. https://doi.org/10.1080/19476337.2023.2199833.

Bell, J. A. and Whiting, S. J. (2004). Effect of fruit on net acid and urinary calcium excretion in an acute feeding trial of women. *Nutrition* 20(5), 492–493. https://doi.org/10.1016/j.nut.2004.01.015.

Bodner, M. and Scampicchio, M. (2020). Does bruising influence the volatile profile of pears? *Nutrition and Food Science* 51(4), 643–652. https://doi.org/10.1108/NFS-05-2020-0213/FULL/PDF.

Boespflug, E. L., Eliassen, J. C., Dudley, J. A., Shidler, M. D., Kalt, W., Summer, S. S., Stein, A. L., Stover, A. N. and Krikorian, R. (2018). Enhanced neural activation with blueberry supplementation in mild cognitive impairment. *Nutritional Neuroscience* 21(4), 297–305. https://doi.org/10.1080/1028415X.2017.1287833.

Boyer, J. and Liu, R. H. (2004). Apple phytochemicals and their health benefits. *Nutrition Journal* 3(1), 5. https://doi.org/10.1186/1475-2891-3-5.

Butkeviciute, A., Viskelis, J., Liaudanskas, M., Viskelis, P. and Janulis, V. (2022). Impact of storage controlled atmosphere on the apple phenolic acids, flavonoids, and anthocyanins and antioxidant activity *in vitro* 11(2), 201. https://doi.org/10.3390/PLANTS11020201.

Cebulj, A., Populin, F., Masuero, D., Vrhovsek, U., Angeli, L., Morozova, K., Scampicchio, M., Costa, F. and Busatto, N. (2023). A multifaced approach sheds light on the molecular details underlaying the mechanism preventing enzymatic browning in 'Majda' apple cultivar (*Malus domestica* Borkh.). *Scientia Horticulturae* 318, 112137. https://doi.org/10.1016/J.SCIENTA.2023.112137.

Ceymann, M., Arrigoni, E., Schärer, H., Bozzi Nising, A. and Hurrell, R. F. (2012). Identification of apples rich in health-promoting flavan-3-ols and phenolic acids by measuring the polyphenol profile. *Journal of Food Composition and Analysis* 26(1–2), 128–135. https://doi.org/10.1016/J.JFCA.2011.12.002.

Chen, H. Y., Jiang, L. F., Zeng, J. H., Huo, Y. R. and Li, Y. X. (2020). Combination of carnauba wax-based coating and 1-methylcyclopropene (1-MCP) maintains better "Fuji" apple qualities during storage at low temperature. *Journal of Food Processing and Preservation* 44(12). https://doi.org/10.1111/JFPP.14925.

Chen, L., Wang, M., Wang, H., Zhou, C., Yuan, J., Li, X. and Pan, Y. (2023). Isothermal storage delays the senescence of post-harvest apple fruit through the regulation of antioxidant activity and energy metabolism. *Foods* 12(9), 1765. https://doi.org/10.3390/FOODS12091765.

Commisso, M., Bianconi, M., Poletti, S., Negri, S., Munari, F., Ceoldo, S. and Guzzo, F. (2021). Metabolomic profiling and antioxidant activity of fruits representing diverse apple and pear cultivars. *Biology* 10(5). https://doi.org/10.3390/biology10050380.

Conforti, F. D. and Totty, J. A. (2007). Effect of three lipid/hydrocolloid coatings on shelf-life stability of golden delicious apples. *International Journal of Food Science and Technology* 42(9), 1101–1106. https://doi.org/10.1111/J.1365-2621.2006.01365.X.

de la Rosa, L. A., Moreno-Escamilla, J. O., Rodrigo-García, J. and Alvarez-Parrilla, E. (2019). Phenolic compounds. *Postharvest Physiology and Biochemistry of Fruits and Vegetables*, 253–271. https://doi.org/10.1016/B978-0-12-813278-4.00012-9.

De León-Zapata, M. A., Sáenz-Galindo, A., Rojas-Molina, R., Rodríguez-Herrera, R., Jasso-Cantú, D. and Aguilar, C. N. (2015). Edible candelilla wax coating with fermented extract of tarbush improves the shelf life and quality of apples. *Food Packaging and Shelf Life* 3, 70–75. https://doi.org/10.1016/J.FPSL.2015.01.001.

Deyman, K. L., Brikis, C. J., Bozzo, G. G. and Shelp, B. J. (2014). Impact of 1-methylcyclopropene and controlled atmosphere storage on polyamine and 4-aminobutyrate levels in "Empire" apple fruit. *Frontiers in Plant Science* 5(APR). https://doi.org/10.3389/FPLS.2014.00144.

Didier, A. J., Stiene, J., Fang, L., Watkins, D., Dworkin, L. D. and Creeden, J. F. (2023). Antioxidant and anti-tumor effects of dietary vitamins A, C, and E. *Antioxidants* 12(3), 632. https://doi.org/10.3390/ANTIOX12030632.

Dixon, J. and Hewett, E. W. (2010). Factors affecting apple aroma/flavour volatile concentration: a review. *New Zealand Journal of Crop and Horticultural Science* 28(3), 155–173. https://doi.org/10.1080/01140671.2000.9514136.

Dresen, E., Lee, Z. Y., Hill, A., Notz, Q., Patel, J. J. and Stoppe, C. (2023). History of scurvy and use of vitamin C in critical illness: a narrative review. *Nutrition in Clinical Practice* 38(1), 46–54. https://doi.org/10.1002/NCP.10914.

Duda-Chodak, A., Tarko, T. and Tuszyński, T. (2011). Antioxidant activity of apples: an impact of maturity stage and fruit part. *Acta Scientiarum Polonorum, Technologia Alimentaria* 10(4), 443–454.

Eran Nagar, E., Okun, Z. and Shpigelman, A. (2020). Digestive fate of polyphenols: updated view of the influence of chemical structure and the presence of cell wall material. *Current Opinion in Food Science* 31, 38–46. https://doi.org/10.1016/J.COFS.2019.10.009.

Espino-Díaz, M., Sepúlveda, D. R., González-Aguilar, G. and Olivas, G. I. (2016). Biochemistry of apple aroma: a review. *Food Technology and Biotechnology* 54(4), 375. https://doi.org/10.17113/FTB.54.04.16.4248.

Fang, T., Cai, Y., Yang, Q., Ogutu, C. O., Liao, L. and Han, Y. (2020a). Analysis of sorbitol content variation in wild and cultivated apples. *Journal of the Science of Food and Agriculture* 100(1), 139–144. https://doi.org/10.1002/jsfa.10005.

Fang, X., An, Y., Zheng, J., Shangguan, L. and Wang, L. (2020b). Genome-wide identification and comparative analysis of GST gene family in apple (*Malus domestica*) and their expressions under ALA treatment. *3 Biotech* 10(7), 307. https://doi.org/10.1007/S13205-020-02299-X.

Feng, S., Yi, J., Li, X., Wu, X., Zhao, Y., Ma, Y. and Bi, J. (2021). Systematic review of phenolic compounds in apple fruits: compositions, distribution, absorption, metabolism, and processing stability. *Journal of Agricultural and Food Chemistry* 69(1), 7-27. Available at: https://pubs.acs.org/doi/full/10.1021/acs.jafc.0c05481.

Ferrentino, G., Giampiccolo, S., Morozova, K., Haman, N., Spilimbergo, S. and Scampicchio, M. (2020a). Supercritical fluid extraction of oils from apple seeds: process optimization, chemical characterization and comparison with a conventional solvent extraction. *Innovative Food Science and Emerging Technologies* 64, 102428. https://doi.org/10.1016/J.IFSET.2020.102428.

Ferrentino, G., Morozova, K., Horn, C. and Scampicchio, M. (2020b). Extraction of essential oils from medicinal plants and their utilization as food antioxidants. *Current Pharmaceutical Design* 26(5), 519–541. https://doi.org/10.2174/1381612826666200121092018.

Francini, A. and Sebastiani, L. (2013). Phenolic compounds in apple (Malus x domestica Borkh.): compounds characterization and stability during postharvest and after processing. *Antioxidants* 2(3), 181–193. https://doi.org/10.3390/ANTIOX2030181.

Gallus, S., Talamini, R., Giacosa, A., Montella, M., Ramazzotti, V., Franceschi, S., Negri, E. and La Vecchia, C. (2005). Does an apple a day keep the oncologist away? *Annals of Oncology* 16(11), 1841–1844.

Geleta, B. T. and Heo, J. Y. (2023). Variation of fruit characteristics and antioxidant properties among ten apple cultivars grown in two different high-latitude regions of South Korea. *Erwerbs-Obstbau*. https://doi.org/10.1007/S10341-023-00934-4.

Gonçalves, J., Ramos, R., Luís, Â., Rocha, S., Rosado, T., Gallardo, E. and Duarte, A. P. (2019). Assessment of the bioaccessibility and bioavailability of the phenolic compounds of *Prunus avium* L. by in vitro digestion and cell model. *ACS Omega* 4(4), 7605–7613. https://doi.org/10.1021/acsomega.8b03499.

Graziani, G., D'Argenio, G., Tuccillo, C., Loguercio, C., Ritieni, A., Morisco, F., Del Vecchio Blanco, C., Fogliano, V. and Romano, M. (2005). Apple polyphenol extracts prevent damage to human gastric epithelial cells in vitro and to rat gastric mucosa in vivo. *Gut* 54(2), 193–200. https://doi.org/10.1136/GUT.2004.046292.

Gumul, D., Ziobro, R., Korus, J. and Kruczek, M. (2021). Apple pomace as a source of bioactive polyphenol compounds in gluten-free breads. *Antioxidants* 10(5). https://doi.org/10.3390/ANTIOX10050807.

Gupta, A. K., Savopoulos, C. G., Ahuja, J. and Hatzitolios, A. I. (2011). Role of phytosterols in lipid-lowering: current perspectives. *QJM* 104(4), 301–308. https://doi.org/10.1093/QJMED/HCR007.

Halliwell, B., Adhikary, A., Dingfelder, M. and Dizdaroglu, M. (2021). Hydroxyl radical is a significant player in oxidative DNA damage in vivo. *Chemical Society Reviews* 50(15), 8355. https://doi.org/10.1039/D1CS00044F.

Hansen, L., Dragsted, L. O., Olsen, A., Christensen, J., Tjønneland, A., Schmidt, E. B. and Overvad, K. (2010). Fruit and vegetable intake and risk of acute coronary syndrome. *British Journal of Nutrition* 104(2), 248–255. https://doi.org/10.1017/S0007114510000462.

Haque, M. A., Morozova, K., Ferrentino, G. and Scampicchio, M. (2021). Electrochemical methods to evaluate the antioxidant activity and capacity of foods: a review. *Electroanalysis* 33(6), 1419–1435. https://doi.org/10.1002/elan.202060600.

Hodgson, J. M., Prince, R. L., Woodman, R. J., Bondonno, C. P., Ivey, K. L., Bondonno, N., Rimm, E. B., Ward, N. C., Croft, K. D. and Lewis, J. R. (2016). Apple intake is inversely associated with all-cause and disease-specific mortality in elderly women 115(5), 860–867. https://doi.org/10.1017/S0007114515005231.

Huang, Y., Qin, S., He, J. and Lyu, D. (2023). Integration of cell wall fraction, organic matter content, and membrane to understand crispness changes in apples. *Scientia Horticulturae* 321, 112309. https://doi.org/10.1016/J.SCIENTA.2023.112309.

Hyson, D. A. (2011). A comprehensive review of apples and apple components and their relationship to human health. *Advances in Nutrition* 2(5), 408–420. https://doi.org/10.3945/AN.111.000513.

Iqbal, S., Tirpanalan-Staben, Ö. and Franke, K. (2022). Modification of dietary fibers to valorize the by-products of cereal, fruit and vegetable industry—a review on treatment methods. *Plants* 11(24). https://doi.org/10.3390/PLANTS11243466.

Jiang, Q. (2017). Natural forms of vitamin E as effective agents for cancer prevention and therapy. *Advances in Nutrition* 8(6), 850–867. https://doi.org/10.3945/AN.117.016329.

Joshipura, K. J., Hu, F. B., Manson, J. E., Stampfer, M. J., Rimm, E. B., Speizer, F. E., Colditz, G., Ascherio, A., Rosner, B., Spiegelman, D. and Willett, W. C. (2001). The effect of fruit and vegetable intake on risk for coronary heart disease. *Annals of Internal Medicine* 134(12). https://doi.org/10.7326/0003-4819-134-12-200106190-00010.

Jung, S. K. and Choi, H. S. (2021). Fruit quality and antioxidant activities of yellow-skinned apple cultivars coated with natural sucrose monoesters. *Sustainability* 13(5), 2423. https://doi.org/10.3390/SU13052423.

Kamrul, H. S. M. M., Schiraldi, A., Cosio, M. S. and Scampicchio, M. (2016). Food and ascorbic scavengers of hydrogen peroxide: A reaction calorimetry investigation. *Journal of Thermal Analysis and Calorimetry* 125(2), 729–737. https://doi.org/10.1007/s10973-015-5170-3.

Kolniak-Ostek, J., Wojdyło, A., Markowski, J. and Siucińska, K. (2014). 1-methylcyclopropene postharvest treatment and their effect on apple quality during long-term storage time. *European Food Research and Technology* 239(4), 603–612. https://doi.org/10.1007/S00217-014-2256-2.

Kuiper, C. and Vissers, M. C. M. (2014). Ascorbate as a co-factor for Fe- and 2-oxoglutarate dependent dioxygenases: physiological activity in tumor growth and progression. *Frontiers in Oncology* 4(NOV). https://doi.org/10.3389/FONC.2014.00359.

Kumari, S., Manohar, S., Kumari, P., Krishnan, V., Maheshwari, C., Narwal, S., Gupta, O. P., Gowda, V. T., Bansal, N., Dahuja, A., Kumari, S., Manohar, S., Kumari, P., Krishnan, V., Maheshwari, C., Narwal, S., Gupta, O. P., Gowda, V. T., Bansal, N. and Dahuja, A. (2023). The role of major phenolics in apple to total antioxidant capacity. *Apple Cultivation - Recent Advances [Working Title]*. https://doi.org/10.5772/INTECHOPEN.109064.

Larsson, S. C., Virtamo, J. and Wolk, A. (2013). Total and specific fruit and vegetable consumption and risk of stroke: a prospective study. *Atherosclerosis* 227(1), 147–152. https://doi.org/10.1016/J.ATHEROSCLEROSIS.2012.12.022.

Łata, B., Trampczynska, A. and Paczesna, J. (2009). Cultivar variation in apple peel and whole fruit phenolic composition. *Scientia Horticulturae* 121(2), 176–181. https://doi.org/10.1016/J.SCIENTA.2009.01.038.

Łata, B., Żakowska-Biemans, S. and Wrona, D. (2022). Apple antioxidant properties as an effect of N dose and rate—mycorrhization involvement: a long-term study. *Antioxidants* 11(12). https://doi.org/10.3390/ANTIOX11122446.

Lee, J., Cheng, L., Rudell, D. R. and Watkins, C. B. (2012). Antioxidant metabolism of 1-methylcyclopropene (1-MCP) treated "Empire" apples during controlled atmosphere storage. *Postharvest Biology and Technology* 65, 79–91. https://doi.org/10.1016/J.POSTHARVBIO.2011.11.003.

Lee, J., Mattheis, J. P. and Rudell, D. R. (2019). High storage humidity affects fruit quality attributes and incidence of fruit cracking in cold-stored 'royal gala' apples. *HortScience* 54(1), 149–154. https://doi.org/10.21273/HORTSCI13406-18.

Li, S., Fasipe, B. and Laher, I. (2022). Potential harms of supplementation with high doses of antioxidants in athletes. *Journal of Exercise Science and Fitness* 20(4), 269. https://doi.org/10.1016/J.JESF.2022.06.001.

Liu, D., Lopez-Sanchez, P. and Gidley, M. J. (2019). Cellular barriers in apple tissue regulate polyphenol release under different food processing and: in vitro digestion conditions. *Food and Function* 10(5), 3008–3017. https://doi.org/10.1039/C8FO02528B.

Maringgal, B., Hashim, N., Mohamed Amin Tawakkal, I. S. and Muda Mohamed, M. T. (2020). Recent advance in edible coating and its effect on fresh/fresh-cut fruits quality. *Trends in Food Science and Technology* 96, 253–267. https://doi.org/10.1016/J.TIFS.2019.12.024.

Meitha, K., Pramesti, Y. and Suhandono, S. (2020). Reactive oxygen species and antioxidants in postharvest vegetables and fruits. *International Journal of Food Science* 2020. https://doi.org/10.1155/2020/8817778.

Millán-Laleona, A., Bielsa, F. J., Aranda-Cañada, E., Gómez-Rincón, C., Errea, P. and López, V. (2023). Antioxidant, antidiabetic, and anti-obesity properties of apple pulp extracts (*Malus domestica* Bork): a comparative study of 15 local and commercial cultivars from Spain. *Biology* 12(7), 891. https://doi.org/10.3390/biology12070891.

Mosibo, O. K., Scampicchio, M. and Ferrentino, G. (2022). Calorimetric adaptation of the inhibited autoxidation method to determine the activity of individual antioxidants and natural extracts. *Journal of Thermal Analysis and Calorimetry* 147(22), 12829–12836. https://doi.org/10.1007/s10973-022-11399-0.

Nakamura, Y., Watanabe, S., Miyake, N., Kohno, H. and Osawa, T. (2003). Dihydrochalcones: evaluation as novel radical scavenging antioxidants. *Journal of Agricultural and Food Chemistry* 51(11), 3309–3312. https://doi.org/10.1021/jf0341060.

Perez-Gago, M. B., Serra, M. and Río, M. A. D. (2006). Color change of fresh-cut apples coated with whey protein concentrate-based edible coatings. *Postharvest Biology and Technology* 39(1), 84–92. https://doi.org/10.1016/J.POSTHARVBIO.2005.08.002.

Pieczywek, P. M., Leszczuk, A., Kurzyna-Szklarek, M., Cybulska, J., Jóźwiak, Z., Rutkowski, K. and Zdunek, A. (2023). Apple metabolism under oxidative stress affects plant cell wall structure and mechanical properties. *Scientific Reports* 13(1). https://doi.org/10.1038/S41598-023-40782-6.

Pisoschi, A. M., Pop, A., Iordache, F., Stanca, L., Predoi, G. and Serban, A. I. (2021). Oxidative stress mitigation by antioxidants - an overview on their chemistry and influences on health status. *European Journal of Medicinal Chemistry* 209, 112891. https://doi.org/10.1016/J.EJMECH.2020.112891.

Platzer, M., Kiese, S., Tybussek, T., Herfellner, T., Schneider, F., Schweiggert-Weisz, U. and Eisner, P. (2022). Radical scavenging mechanisms of phenolic compounds: a quantitative structure-property relationship (QSPR) study. *Frontiers in Nutrition* 9, 882458. https://doi.org/10.3389/fnut.2022.882458.

Prange, R. K. and Wright, A. H. (2023). A review of storage temperature recommendations for apples and pears 12(3), 466. https://doi.org/10.3390/FOODS12030466.

Preedy, V. R. and Patel, V. B. (2022). Understanding and optimising the nutraceutical properties of fruit and vegetables. Available at: https://shop.bdspublishing.com/store/bds/detail/workgroup/3-190-109523.

Preti, R. and Tarola, A. M. (2021). Study of polyphenols, antioxidant capacity and minerals for the valorisation of ancient apple cultivars from Northeast Italy. *European Food Research and Technology* 247(1), 273–283. Available at: https://link.springer.com/article/10.1007/s00217-020-03624-7.

Rice-Evans, C. A., Miller, N. J. and Paganga, G. (1996). Structure-antioxidant activity relationships of flavonoids and phenolic acids. *Free Radical Biology and Medicine* 20(7), 933–956. https://doi.org/10.1016/0891-5849(95)02227-9.

Sadgrove, N. J. and Simmonds, M. S. J. (2022). Advances in understanding the role of plant phytochemicals in preventing cardiovascular disease. In: *Understanding and Optimising the Nutraceutical Properties of Fruit and Vegetables*, pp. 235–270. https://doi.org/10.19103/AS.2022.0101.09.

Scampicchio, M., Wang, J., Blasco, A. J., Sanchez Arribas, A., Mannino, S. and Escarpa, A. (2006). Nanoparticle-based assays of antioxidant activity. *Analytical Chemistry* 78(6), 2060–2063. https://doi.org/10.1021/ac052007a.

Serra, A. T., Rocha, J., Sepodes, B., Matias, A. A., Feliciano, R. P., De Carvalho, A., Bronze, M. R., Duarte, C. M. M. and Figueira, M. E. (2012). Evaluation of cardiovascular protective effect of different apple varieties - correlation of response with composition. *Food Chemistry* 135(4), 2378–2386. https://doi.org/10.1016/J.FOODCHEM.2012.07.067.

Serra, S., Anthony, B., Boscolo Sesillo, F., Masia, A. and Musacchi, S. (2021). Determination of post-harvest biochemical composition, enzymatic activities, and oxidative browning in 14 apple cultivars. *Foods* 10(1). https://doi.org/10.3390/FOODS10010186.

Shen, N., Wang, T., Gan, Q., Liu, S., Wang, L. and Jin, B. (2022). Plant flavonoids: classification, distribution, biosynthesis, and antioxidant activity. *Food Chemistry* 383. https://doi.org/10.1016/J.FOODCHEM.2022.132531.

Shen, X., Su, Y., Hua, Z., Sheng, L., Mendoza, M., He, Y., Green, T., Hanrahan, I., Blakey, R. and Zhu, M. J. (2021). Effectiveness of low-dose continuous gaseous ozone in controlling *Listeria innocua* on red delicious apples during 9-month commercial cold storage. *Frontiers in Microbiology* 12. https://doi.org/10.3389/FMICB.2021.712757.

Song, Y., Manson, J. E., Buring, J. E., Sesso, H. D. and Liu, S. (2005). Associations of dietary flavonoids with risk of type 2 diabetes, and markers of insulin resistance and systemic inflammation in women: a prospective study and cross-sectional analysis. *Journal of the American College of Nutrition* 24(5), 376–384. https://doi.org/10.1080/07315724.2005.10719488.

Su, G., Qin, X., Yang, C., Sabatino, A., Kelly, J. T., Avesani, C. M. and Carrero, J. J. (2022). Fiber intake and health in people with chronic kidney disease. *Clinical Kidney Journal* 15(2), 213–225. https://doi.org/10.1093/CKJ/SFAB169.

Suhag, R., Ferrentino, G., Morozova, K., Zatelli, D., Scampicchio, M. and Amorati, R. (2024). Antioxidant efficiency and oxidizability of mayonnaise by oximetry and isothermal calorimetry. *Food Chemistry* 433. https://doi.org/10.1016/J.FOODCHEM.2023.137274.

Tahir, H. E., Xiaobo, Z., Mahunu, G. K., Arslan, M., Abdalhai, M. and Zhihua, L. (2019). Recent developments in gum edible coating applications for fruits and vegetables preservation: a review. *Carbohydrate Polymers* 224. https://doi.org/10.1016/J.CARBPOL.2019.115141.

Treutter, D. (2001). Biosynthesis of phenolic compounds and its regulation in apple. *Plant Growth Regulation* 34(1), 71–89. Available at: https://link.springer.com/article/10.1023/A:1013378702940.

Van Der Sluis, A. A., Dekker, M. and Van Boekel, M. A. J. S. (2005). Activity and concentration of polyphenolic antioxidants in apple juice. 3. Stability during storage. *Journal of Agricultural and Food Chemistry* 53(4), 1073–1080. https://doi.org/10.1021/JF040270R.

Velderrain-Rodríguez, G. R., Palafox-Carlos, H., Wall-Medrano, A., Ayala-Zavala, J. F., Chen, C.-Y. Y. O. O., Robles-Sánchez, M., Astiazaran-García, H., Alvarez-Parrilla, E. and González-Aguilar, G. A. (2014). Phenolic compounds: their journey after intake. *Food and Function* 5(2), 189–197. https://doi.org/10.1039/C3FO60361J.

Vrhovsek, U., Rigo, A., Tonon, D. and Mattivi, F. (2004). Quantitation of polyphenols in different apple varieties. *Journal of Agricultural and Food Chemistry* 52(21), 6532–6538. https://doi.org/10.1021/JF049317Z.

Watkins, C. B. (2006). The use of 1-methylcyclopropene (1-MCP) on fruits and vegetables. *Biotechnology Advances* 24(4), 389–409. https://doi.org/10.1016/J.BIOTECHADV.2006.01.005.

Watkins, C. B., Nock, J. F. and Whitaker, B. D. (2000). Responses of early, mid and late season apple cultivars to postharvest application of 1-methylcyclopropene (1-MCP) under air and controlled atmosphere storage conditions. *Postharvest Biology and Technology* 19(1), 17–32. https://doi.org/10.1016/S0925-5214(00)00070-3.

Watkins, C. B., Silsby, K. J. and Goffinet, M. C. (1997). Controlled atmosphere and antioxidant effects on external CO_2 injury of "Empire" apples. *HortScience* 32(7), 1242–1246. https://doi.org/10.21273/HORTSCI.32.7.1242.

Whitaker, B. D. (2004). Oxidative stress and superficial scald of apple fruit. *HortScience* 39(5), 933–937. https://doi.org/10.21273/HORTSCI.39.5.933.

Williamson, G. (2017). The role of polyphenols in modern nutrition. *Nutrition Bulletin* 42(3), 226–235. https://doi.org/10.1111/NBU.12278.

Wu, X. and Prior, R. L. (2005). Systematic identification and characterization of anthocyanins by HPLC-ESI-MS/MS in common foods in the United States: fruits and berries. *Journal of Agricultural and Food Chemistry* 53(7), 2589–2599. https://doi.org/10.1021/jf048068b.

Yin, X., Chen, K., Cheng, H., Chen, X., Feng, S., Song, Y. and Liang, L. (2022). Chemical stability of ascorbic acid integrated into commercial products: a review on bioactivity and delivery technology. *Antioxidants* 11(1). https://doi.org/10.3390/ANTIOX11010153.

Zhang, W., Li, X. and Jiang, W. (2020a). Development of antioxidant chitosan film with banana peels extract and its application as coating in maintaining the storage quality of apple. *International Journal of Biological Macromolecules* 154, 1205–1214. https://doi.org/10.1016/J.IJBIOMAC.2019.10.275.

Zhang, X., Xu, J., Xu, Z., Sun, X., Zhu, J. and Zhang, Y. (2020b). Analysis of antioxidant activity and flavonoids metabolites in peel and flesh of red-fleshed apple varieties. *Molecules* 25(8). https://doi.org/10.3390/molecules25081968.

Zheng, H. Z., Kim, Y. I. and Chung, S. K. (2012). A profile of physicochemical and antioxidant changes during fruit growth for the utilisation of unripe apples. *Food Chemistry* 131(1), 106–110. https://doi.org/10.1016/J.FOODCHEM.2011.08.038.

Part 2

Breeding and crop management to optimise quality

Chapter 5

Breeding for fruit quality improvement in apple

Soon Li Teh[1], *Washington State University, USA; Sarah Kostick*[2], *University of Minnesota, USA; and Kate Evans*[1], *Washington State University, USA*

1 Introduction

Domesticated apple (*Malus domestica* Borkh; family Rosaceae) is one of the most economically valuable temperate fruit crops. In 2020, global apple production was 86 million tons with China, USA, Turkey, Poland, and India being the top five producing countries (FAOSTAT, 2020). Most apple species are diploid ($2n = 2x = 34$) with ancestral haploid origin of $x = 9$ (Velasco et al., 2010). The high heterozygosity of an outcrosser and ancient autotetraploid origin render genome assembly and genetic studies challenging (Zharkikh et al., 2008; Velasco et al., 2010).

Apple breeding is a long-term and resource-intensive process. Apple has a juvenility phase of 5 to 10 years, resulting in a long generation time. The time and cost elements are further compounded by considerable land area that is required to grow and maintain seedling trees. However, developing improved apple scion varieties that meet consumer and stakeholder expectations is vital for economic sustainability of the apple industry. Breeding programs must continue to target traits of high value to all entities in the supply chain (e.g. consumers, market intermediaries, producers).

[1] Washington State University, Tree Fruit Research and Extension Center, Wenatchee, WA 98801, USA.
[2] University of Minnesota, Department of Horticultural Science, Saint Paul, MN 55108, USA.

http://dx.doi.org/10.19103/AS.2023.0127.07

Fruit quality is an overarching breeding target that includes appearance (e.g. color, shape, size), eating quality (e.g. acidity, flavors, sweetness, texture, volatiles), and storability (shelf life and resistance to disorders) (Shewfelt, 1999). Storability is important as most apples are stored for extended periods of time for year-round availability to consumers. Resistance to diseases (e.g. apple scab, fire blight, powdery mildew) and pests are also major targets in breeding programs. Depending on the program, additional targets may include cold hardiness, tolerance to abiotic stresses, and regular cropping. In this chapter, the Washington State University apple breeding program (WABP) schema is used as an example. While every program may differ in specific targets and methodologies, most follow a similar breeding scheme. The WABP schema consists of four stages: seedling production, Phase 1, Phase 2, and Phase 3 (Evans, 2013). Seedlings are generated via controlled hybridization. DNA-informed breeding can be implemented in this stage to choose breeding parents, and assist in selecting or culling seedlings, thereby reducing the need to maintain unfavorable trees to fruit set. In Phase 1, fruit is assessed for the first time from selected seedlings. In Phase 2, extensive data is collected on seedlings selected in Phase 1 that are replicated in small multi-orchard trials. In Phase 3, selections advanced from Phase 2 are replicated further and evaluated for commercial viability.

Developing apple varieties based on genetic knowledge and markers linked to priority breeding traits, where numerous favorable attributes can be targeted, may provide a solution for sustainable apple breeding. The foundational step is to elucidate genetic control of quality traits. Subsequently, translational steps are required to convert genomic outputs into practical tools for DNA-informed breeding. DNA-informed breeding is defined as the use of DNA information, obtained through direct assays of a plant's DNA, to guide breeding decisions (Peace, 2017). It encompasses trait-predictive diagnostics, parentage ascertainment, genetic fingerprinting, marker-assisted seedling selection, as well as various other breeding applications, as reviewed by Peace (2017). DNA-informed breeding provides several advantages, including potential long-term cost savings (Edge-Garza et al., 2015; Wannemuehler et al., 2019), pyramiding of favorable alleles, informed choice of parents, early generation selection of seedlings, and ultimately increased breeding efficiency and precision (Collard and Mackill, 2008; Xu and Crouch, 2008; Kellerhals et al., 2009; Peace, 2017).

Increased awareness of genomic and breeding challenges culminated in interdisciplinary projects that focused on developing, converting, and evaluating tools to increase the efficiency of apple-breeding process. Strategic collaborations included High-quality Disease Resistant Apples for a Sustainable Agriculture (HiDRAS; https://sites.unimi.it/camelot/hidras/; Gianfranceschi and Soglio, 2004), FruitBreedomics (http://fruitbreedomics.com/; Laurens et al., 2010), RosBREED and RosBREED2 (https://www.rosbreed.org/; Iezzoni et al., 2010, 2017).

However, there has been a disconnect between genomic discoveries and their applications toward variety development, described as 'application gap' by Collard and Mackill (2008). Fulfillment of DNA-informed breeding potential has been hampered by lack of predictive DNA tests relevant to breeding germplasm, as well as access to cost-effective and timely diagnostic services (Ru et al., 2015; Evans et al., 2012; Evans and Peace, 2017; Peace, 2017). In recent years, locus-specific DNA-predictive tests encompassing various fruit quality, biotic resistance, and tree physiological traits have become available for routine DNA-informed breeding, as summarized by Evans and Peace (2017).

This chapter provides an update to Iwanami (2011) with a review of apple quality trait phenotyping and its challenges, a framework for DNA-informed breeding, as well as how quality traits are being incorporated into apple breeding programs using molecular tools.

2 Phenotyping

Phenotyping fruit quality for routine selection in breeding programs is challenging because many traits are quantitatively controlled and strongly influenced by environmental factors as well as horticultural/management practices (Iwanami, 2011; Brown, 2012). Variability in light intensity and duration, day and night temperatures, humidity, precipitation, and soil characteristics are environmental factors that can impact the final quality of the fruit (Corelli-Grappadelli and Lakso, 2004; Musacchi and Serra, 2018). In addition, fruit quality is influenced by horticultural factors, such as canopy management, planting density, choice of rootstocks, crop load management, and irrigation practices (Wagenmakers and Callesen, 1995; Serra et al., 2016; Musacchi and Serra, 2018; Reid and Kalcsits, 2020).

Different combinations of environmental and horticultural factors hamper standardization of fruit sampling for phenotyping, which is further complicated by trait subjectivity (e.g. estimating fruit maturity), use of indirect measures (e.g. estimating sweetness using soluble solids concentrations (SSC)), and applicability of metrics across samples (e.g. starch-iodine indices of different varieties) (Fig. 1). In addition, fruit maturity at harvest is a major factor influencing storability. Watkins (2017) described the quality-storability tradeoff, where fruit harvested earlier within an accession's harvest window (i.e. less mature) tend to have longer storage potential, but exhibit lower quality characteristics. In published works and across breeding programs, fruit is phenotyped/evaluated at harvest (i.e. shortly after harvest), or after being stored in refrigerated coolers of varying temperatures (e.g. 1–4°C), and of different storage conditions (i.e. regular versus controlled atmosphere) for variable durations (e.g. 2, 4 or 6 months), which subsequently may or may not be subjected to a warming/

Figure 1 Fruit maturity is highly influenced by environmental and horticultural factors, as evidenced by the within-sample variability of starch-iodine pattern indices. Each row of five fruit belongs to the same accession that were harvested on the same date.

Figure 2 Sensory assessment of fruit quality (predominantly eating quality) in Washington State University apple breeding program, where data are collected on electronic tablets.

equilibrium period prior to phenotyping. In summary, lack of standardized phenotyping protocols, large within-sample variations, and differences in genetic materials impede direct comparisons and knowledge transfer across programs (Teh et al., 2021b).

Fruit quality is typically phenotyped destructively and can be divided into: (1) instrumental measurements and (2) sensory assessment (Fig. 2). Instrumental measurements of fruit quality traits include firmness and crispness via penetrometers or texture analyzers, SSC via refractometers, and acidity via titrators. Sensory assessment is described here as an assessment of fruit sensory attributes (e.g. acidity, crispness, firmness, juiciness, sweetness) by breeding team member(s) on a hedonic scale from weakest to strongest perceptions (Evans et al., 2010; Teh et al., 2020a; Teh et al., 2021a) (Fig. 2). This term is contrasted with sensory analysis/panel, where trained panelists rate samples based on a calibrated scale. This terminology distinction, while critical, does not diminish the accuracy of sensory assessment of a breeding team that routinely tastes up to 100 samples at a time to meet the necessary throughput of the program (Teh et al., 2020a). In a breeding program where fruit of diverse seedlings/selections are routinely tasted and instrumentally measured, sensory attributes and the corresponding instrumental metrics were correlated with moderate to high coefficients (Evans et al., 2010; Teh et al., 2020a; Teh et al., 2021a). In addition, fruit appearance (e.g. size, shape, color coverage, lenticels) as well as external and internal disorders are traits rated subjectively (e.g. severity of russet, severity of scald).

Non-destructive phenotyping technologies are becoming increasingly important in commercial/industrial applications (e.g. fruit sorting) due to their improved efficiencies and reduced reliance on labor (e.g. costs and subjectivity). Non-destructive systems based on imaging, spectroscopy, and spectral imaging (i.e. combination of imaging and spectroscopy) are implemented to detect disorders, chemical residues, and quality indices, as summarized by Tan et al. (2019). These tools were originally developed for commercial applications, but may offer great utility to breeding programs, which evaluate large numbers of accessions with often limited fruit, especially during early stages of selection. These technologies need to be adapted to analyze/predict diverse accessions since most industrial applications focus on single-variety trait prediction.

A precursory requirement for non-destructive trait prediction is to construct a statistical model where a non-destructive metric is highly correlated with a destructive measure. Virtually all non-destructive tools rely on vision imaging at various wavelengths, such as: (1) near infrared (NIR) at 750–2500 nm, (2) hyperspectral at 400–1100 nm, and (3) subtractive (e.g. delta absorbance at 670 nm and 720 nm, or $I_{AD} = A_{670nm} - A_{720nm}$). Destructively measured traits that have been reliably predicted with corresponding non-destructive metrics include dry matter concentration (with NIR), SSC (with NIR), fruit bruises (with hyperspectral), and chlorophyll content of fruit skin (with I_{AD}), as previously reviewed by Teh et al. (2021b). However, non-destructive models were largely tested on variety-specific predictions yielding higher accuracies than can be expected from predicting diverse accessions in a breeding program (Zhang et al., 2019; Teh et al., 2020b). The most common non-destructive metric is near infrared (NIR) spectroscopy; concepts, strategies, and implementations were extensively described by Nicolai et al. (2007), Porep et al. (2015), and Wang et al. (2015). Successful application of non-destructive phenotyping in breeding programs can be hampered by large within-sample variance owing to diverse genetic materials being evaluated. For example, NIR prediction accuracy of dry matter concentrations in WABP was impacted by year, orchard site, and tree age (Teh et al., 2020b).

In the ensuing fruit quality section, use of phenotypic selection (with and without DNA-based tools) to incorporate quality traits in apple scion breeding programs will be described.

3 DNA-informed breeding

Advancements in genomics research and DNA technologies have resulted in a wealth of genomics outputs. The first apple draft genome sequence (Velasco et al., 2010) paved the way for subsequent progress, such as a high-quality *de novo* assembly of the apple genome (Daccord et al., 2017) and an apple genome variation map (Duan et al., 2017). Two international research consortia, RosBREED and FruitBreedomics, provided bioinformatics and molecular resources to enable the development of apple SNP arrays: 8K (Chagné et al., 2012a), 20K (Bianco et al., 2014), and 480K (Bianco et al., 2016). In addition, genotyping-by-sequencing (GBS) has been a cost-effective alternative to generate a large number of polymorphic markers for genetic mapping (Gardner et al., 2014). These genomics advancements, summarized as draft genome sequences and genotyping platforms, have been a major impetus to downstream outputs of genetic maps, quantitative trait locus (QTL) discoveries, and DNA fingerprinting. While the growing outputs are critical in fulfilling intellectual curiosity, their practical breeding applications (through DNA-informed breeding) have been mostly unrealized (Bliss, 2010). In recent years, there have been considerable efforts to focus on translating genomics resources to breeding applications, which are becoming relatively routine in many apple scion breeding programs. Examples of DNA-informed breeding applications include identifying seedlings with desirable alleles (one or multiple pyramided trait loci), choosing complementary parents, confirming parentage, among others detailed in Peace (2017).

Tools must be adapted for local breeding utility, which necessitates knowledge of trait-locus genomic positions and the conversion into locus-specific, performance-predictive assays relevant to a breeding program's germplasm to distinguish allelic segregation (Peace, 2017). Once a candidate DNA test is selected for a trait locus, allelic effects (i.e. phenotype-genotype association) are systematically validated in different genetic backgrounds (Peace et al., 2014). Marker testing can utilize simple polymerase chain reaction (PCR)-based, or single nucleotide polymorphism (SNP)-based markers. The former is versatile across various genotyping platforms and amenable to multiplexing, while the latter is gaining popularity with ever-increasing cost efficiencies.

In addition, a breeder should identify a cost-efficient application scheme, factoring in monetary resources for routine use and initial establishment

(Luby and Shaw, 2001; Ru et al., 2015; Peace, 2017). In the case of rosaceous crops such as apple, for which high input costs are incurred in multiple years of growing and maintaining seedlings/selections before evaluation, a DNA-informed application scheme is largely deemed cost-efficient (Edge-Garza et al., 2015). Additional efficiency factors should also be considered, such as trait genetic gain (Ru et al., 2016) and break-even point from seedling cull rate (Wannemuehler et al., 2019). Availability of reasonably priced services to obtain DNA test results with reliable timeliness can be major logistical factors in determining feasibility of DNA-informed breeding in a breeding program (Evans and Peace, 2017; Peace, 2017). In a marker-assisted seedling selection process involving tissue collection, DNA extraction, and genotyping, a breeder must receive readily interpretable DNA results in time to keep or cull seedlings before transitioning seedlings to a new location or pot size (Edge-Garza et al., 2010; Evans and Peace, 2017). Dependability and trust relationships with the diagnostics service provider are of paramount importance to the breeding program's success.

4 Fruit quality traits

4.1 Appearance

Fruit appearance encompasses multiple traits, including size, shape, and color coverage (Fig. 3), that collectively influence consumer preference. Fruit appearance traits are often rated subjectively in apple breeding programs. For instance, WABP members use a list of predefined sensory attributes (e.g. size, shape, percent red color coverage, presence of russet) to assess a set of five fruit per accession/selection based on ordinal rating scales (Evans et al., 2010;

Figure 3 Diverse appearance in size, shape, and skin finish.

Teh et al., 2020a). Though most traits can be phenotyped with instrumental tools (e.g. skin color grading with a colorimeter), reliance on instrument measurements removes the 'human perception' elements in judging the complex interplay of fruit quality traits. Most breeding programs implement a dual/hybrid system of instrumental measurements and sensory assessment, where some traits are evaluated with both.

Fruit color is a quality factor affecting consumers' willingness to purchase and market value (King and Cliff, 2002). Red fruit coloration of apple is determined by anthocyanin concentration that is controlled by genetics and environmental conditions. *MdMYB1* and *MdMYBA* transcription factors regulate red pigmentation in the skin (Takos et al., 2006, Ban et al., 2007), and *MdMYB10* in the flesh (Espley et al., 2007). Gene expression for anthocyanin synthesis and accumulation in the skin was higher among varieties (e.g. Fuji) with at least one allele of *MdMYB1-1* than those without (e.g. Golden Delicious: *MdMYB1-2/MdMYB1-3*) (Takos et al., 2006). A predictive DNA test using derived cleaved amplified polymorphism sequence (dCAPS) PCR with *Bst*EII restriction assay was developed to distinguish the presence of the *MdMYB1-1* allele (i.e. 263 bp) versus absence (i.e. 291 bp) (Takos et al., 2006; Zhu et al., 2011; Zhang et al., 2014). An A-allele marker for *MdMYBA* designed by Cheng et al. (1996) yielded three PCR products, with fragment A^1 (1180 bp) being linked to red skin, while a^1 (1230 bp) and a^2 (1320 bp) to yellow skin (Zhang et al., 2014). In addition, a SNP marker, ss475879531 was developed for real-time PCR assay, where genotypes 'GG', 'AG', and 'AA' were associated with high-red, medium-red, and low-red percent (over)color skin phenotypes, respectively (Chagné et al., 2016). This marker is located approximately 878 kb upstream of *MdMYB1*, suggesting a low-to-moderate likelihood for a recombination between them. Subsequently, a microsatellite marker, Mdo.chr9.4 (located 8.3 kb from *MdMYB1*) was developed, where allele R_0 (157 bp) was associated with overcolor intensity in a dose-dependent manner (Moriya et al., 2017a). A DNA predictive test is also available for selection of red-core apples. A 786-bp PCR product from a MdMYB10 insertion-deletion (indel) marker was strongly linked with red-core fruit borne on red foliage trees (Chagné et al., 2007).

Fruit size is another appearance trait that influences consumers' purchasing decision. Kostick and Luby (2022) identified two fruit size (phenotyped as fruit weight) loci from 'Honeycrisp'-derived families on chromosomes 8 and 16 that collectively explained 15% of phenotypic variance. The presence of haplotypes 8C, 16B, 16D, and 16F (i.e. number denotes chromosome) was significantly associated with high fruit weight, while the presence of haplotypes 8A, 8B, 8G, and 16A was linked with low fruit weight (Kostick and Luby, 2022). It is worth noting that the chromosome 16 locus colocalizes with other important trait loci, such as acidity *Ma* and bitter pit susceptibility *Bp-2*.

4.2 Eating quality

Eating quality is a complex term that encompasses fruit textural characteristics (e.g. crispness, firmness, juiciness), interplay of sweetness/sugars and acidity, as well as volatiles. In 2011–12, a US nationwide producer survey was carried out to systematically identify and rank fruit quality and tree traits valued by producers (Yue et al., 2013). The top five discrete traits, in descending order of priority, were fruit flavor, crispness, firmness, shelf life at retail, and juiciness (Yue et al., 2013). Fruit firmness and crispness are also the most valued traits among consumers (Harker et al., 2008).

4.2.1 Texture

In most apple breeding programs, fruit texture is measured instrumentally (to capture mechanical and/or acoustic components), as well as rated subjectively on a hedonic scale. Mechanical texture components are measured via puncture tests using a mechanical/computerized penetrometer to assess firmness, crispness, and other texture traits. For instance, WABP routinely uses a Mohr™ Digi-Test (Mohr and Associates, Richland, WA) penetrometer to estimate fruit texture trait values. Penetrometer texture outputs were shown to correlate well with sensory crispness and firmness (Evans et al., 2010; Teh et al., 2020a). Fruit texture can also be acoustically profiled, which was shown to correlate with sensory crispness (Taniwaki et al., 2006; Costa et al., 2011). Simultaneous profiling of mechanical displacement and acoustic response can be achieved by coupling a texture analyzer with an acoustic envelope device (Costa et al., 2011). In a genetic analysis of apple texture using a high-resolution texture analyzer, mechanical components were mapped on chromosome 10, while acoustic components on chromosomes 2 and 14 (Di Guardo et al., 2017). Corresponding haplotypes for mechanical and acoustic loci were characterized (Di Guardo et al., 2017), but have not been validated in subsequent studies. In addition, the complexity of fruit texture and the use of different instruments (and consequently, different outputs) hamper direct comparisons among studies.

Fruit textural traits were also mapped using sensory traits. In a genomewide association study based on descriptive sensory traits, sensory crispness, juiciness, and mealiness were jointly mapped to the same genomic region on chromosome 15. SNP marker chr13:6,049,060 was predictive in discriminating fruit with unfavorable levels of crispness, juiciness, and mealiness, which were statistically associated with genotype CC (Amyotte et al., 2017). Because fruit texture (e.g. firmness of outer and inner cortices) is related to fruit integrity and postharvest quality, other texture-related genes/loci (e.g. *Md-ACS*, *Md-ACO*, *Md-PG1*, *Md-Exp7*) and their corresponding diagnostic DNA markers are described in *Section 4.3 Storability*.

More recently, 56 candidates genes for flesh firmness retention and crispness retention loci were identified in 3 bi-parental crosses of 2664 individuals with 0.84 broad-sense heritability (Wu et al., 2021). An 8-bp deletion in the *MdERF3* promoter on chromosome 3 and a 3-bp deletion in the *MdERF118* promoter on chromosome 16 were both associated with decreased retention of firmness and crispness. A list of predictive SNP markers with varying range of gene actions (from partial to complete dominance) was provided, but these markers may not be applicable in germplasm lacking *M. asiatica* Nakai, which was a parent ('Zisai Pearl') in two of the three crosses (Wu et al., 2021).

4.2.2 Sweetness

Sweetness, acidity, and the balance between the two are the main drivers for consumer preference and perception of a desirable apple-eating experience (Iwanami, 2011). Sugars in apple are primarily comprised of fructose, sucrose, glucose, and sorbitol (Fuleki et al., 1994), while acidity in a mature fruit is predominated by malic acid (Iwanami, 2011). These levels are heavily influenced by environmental conditions, canopy management, crop load, fruit maturity, and storage conditions. In most apple breeding programs, sweetness is typically estimated as SSC with a refractometer, and rated with sensory assessment. Individual sugars and SSC were mapped on virtually all apple chromosomes, but stable/consistent loci for fructose and sucrose concentrations were mapped on chromosome 1, explaining 34–67% and 13–41% of phenotypic variations, respectively (Guan et al., 2015). In a genomewide association study of 149 varieties, sucrose concentration and percent fructose of total sugars were also mapped to the same genomic region as Guan et al. (2015), explaining 27% and 47% of phenotypic variations, respectively (Larsen et al., 2019). Fruit of genotype CC (of SNP marker chr1:30129468) was associated with elevated levels of sucrose concentrations, or a lower percentage of fructose to total sugars (Larsen et al., 2019). In dissecting candidate genes that regulate sugar accumulation in apple fruit, three *SWEET* (Sugar Will Eventually be Exported Transporters) genes, namely *MdSWEET2e* (on chromosome 10), *MdSWEET9b* (chromosome 4), and *MdSWEET15a* (chromosome 16) were associated with individual and total sugar concentrations of at-harvest fruit (Zhen et al., 2018). However, these genes collectively explained 9.3% of fructose concentrations and 14.6% of total sugar concentrations, thereby indicating the highly quantitative nature of sweetness/sugars trait. Diagnostic DNA markers were developed to screen the presence or absence of these genes, where alleles $(AT)_{13}$, $(CT)_{19}$, and T of *MdSWEET2e*, *MdSWEET9b*, and *MdSWEET15a* markers, respectively, were associated with higher total sugar concentrations (Zhen et al., 2018).

4.2.3 Acidity

Fruit acidity is typically measured through titration of apple juice (i.e. titratable acidity) and can also be rated subjectively through sensory assessment. Two major acidity loci were mapped - *Ma* on chromosome 16 (Maliepaard et al., 1998) and *Ma3* on chromosome 8 (Kenis et al., 2008). Two aluminum-activated malate transporter-like (*MdALMTII*) genes (i.e. *Ma1* and *Ma2*) were suggested as candidate genes of *Ma* (chromosome 16), narrowing down the locus to 65-82kb with 12-19 predicted genes (Bai et al., 2012). *Ma1* co-segregated with fruit acidity, where $CAPS_{1455}$ (cleaved amplified polymorphism sequence) marker can be used to select individuals with low acidity allele, *Ma1-1455A* (Bai et al., 2012).

Subsequently, marker MdALMTII was validated by Jia et al. (2018), where individuals with AA genotype showed significantly low levels of fruit malate content. A new marker targeting *Ma3* (chromosome 8) was also developed. A 184-bp insertion-deletion marker for *MdPP2CH* detected co-segregation of A/G variants with fruit malate content, where individuals with AA genotype had significantly low levels of fruit malate. Additionally, a 36-bp insertion-deletion marker in the *MdSAUR37* promoter region (also targeting *Ma3*) can be used to select for low fruit malate, which was linked with S^P (allele i). This work suggested a hierarchy of genetic interaction, where the *MdSAUR37/MdPP2CH/MdALMTII* regulatory chain cascaded hierarchical epistatic effects on apple fruit malate content (Jia et al., 2018).

In a separate study, *Ma* and *Ma3* were mapped, jointly explaining 665% of phenotypic variation (Verma et al., 2019). Both loci were haplotyped, but careful interpretation is required. For *Ma* haplotypes, B alleles of ss475881815 and ss475882553 markers were associated with the high-acidity allele, also termed Q allele. *Ma3* was haplotyped using nine SNP markers, where four unique haplotypes were associated with Q allele. An additive allele dosage model was proposed for these loci, with an average of +1.8 mg/L malic acid at harvest per high acidity allele (Verma et al., 2019). More recently, *Ma* and *Ma3*, as well as two additional loci (*Ma4* and *Ma5* on chromosomes 6 and 1, respectively) were mapped using sensory evaluation traits (Rymenants et al., 2020). AX-105220211 was suggested as a predictive marker at the *Ma3* locus, to discriminate between Q_{Ma3} (i.e. high acidity associated with the allele A) and q_{Ma3} (i.e. low acidity) alleles, and reconciled inconsistent results from a previous study by Verma et al. (2019).

4.2.4 Volatiles

In addition to acidity and sweetness, another element of flavor is aroma - a complex mixture of volatile compounds detected by olfactory sensations (Iwanami, 2011). Over 300 volatiles of primarily alcohols, aldehydes, and esters were reported in fresh apples (Dixon and Hewett, 2000). Expression and composition of fruit volatiles are strongly influenced by environmental/external conditions during fruit maturity, ripening, postharvest handling, and storage settings (El Hadi et al., 2013). Due to significant environmental variability on fruit volatile composition, standardization of fruit collection for phenotyping is extremely challenging, if not impossible. Furthermore, the volatile/unstable nature of aroma compounds imposes significant logistical constraints to quench or preserve volatile compounds for quantitative analysis (Song and Forney, 2008).

Fifty loci associated with 27 volatiles were mapped on chromosomes 2, 3, and 9 (Dunemann et al., 2009). An alcohol acyl-transferase (AAT) candidate gene was also mapped on chromosome 2, co-localizing with a cluster of loci for several apple fruit esters (Dunemann et al., 2009). *MdAAT1* haplotype was constructed using four SNP markers, where haplotype H8 (i.e. T-G-T-G found in 28 varieties) was associated with normal-to-elevated ester concentrations, while H1 (i.e. C-A-C-A found in 102 varieties) with significantly decreased ester concentrations (Dunemann et al., 2012). Cappellin et al. (2015) developed a microsatellite marker, termed Md-AAT1$_{SSR}$ that was screened on 124 apple accessions, where allele 201 was associated with elevated levels of volatile organic compounds, but significantly decreased levels of acetate esters. In a genomewide association study of 149 varieties, loci for two acetate esters (i.e. butyl acetate and hexyl acetate) were mapped also on chromosome 2 in the hotspot region for genes involved in aroma volatile biosynthesis, such as *MdAAT1* (Larsen et al., 2019). Allele T (of SNP chr2:1258734) and allele A (of SNP chr2:1730413) were associated with significantly elevated levels of butyl- and hexyl acetate, respectively (Larsen et al., 2019).

4.2.5 Polyphenols

Apple fruit contains many polyphenolic compounds, which are implicated in the prevention of degenerative, neurodegenerative, and cardiovascular diseases due to their high antioxidant properties (Arts and Hollman, 2005). Five major groups of polyphenols in apple are anthocyanidins, dihydrochalcones, flavanols, flavonols, and hydroxycinnamic acids (Iwanami, 2011). A cluster of large-effect loci associated with fruit flavanols catechin, epicatechin, procyanidin dimer, and five unknown procyanidin oligomers was mapped on chromosome 16 (Chagné et al., 2012b). These loci co-localized with *Leucoanthocyanidin reductase* (*LAR1*), a candidate gene encoding the enzyme that catalyzes the conversion of leucoanthocyanidins into catechin. A high-resolution melting (HRM)-based marker for *LAR1* was developed to aid in selection for increased flavanol content. In an <*ef* × *eg*> segregation pattern, individuals with homozygous *ee* genotype showed significantly lower flavanol concentrations in fruit skin, and no flavanol in fruit cortex (Chagné et al., 2012b). However, given that this cluster of flavanol loci also co-localizes with loci for malic acid (*Ma*), loss of fruit firmness during storage, susceptibility to bitter pit, and budbreak date, the *LAR1* marker should be used in tandem with predictive markers of other traits to enable selection of progeny with the most desirable combination of traits.

Untargeted metabolic profiling of peel and flesh tissues in a segregating F_1 family yielded 669 metabolic loci that were predominantly clustered as hotspots on chromosomes 1, 8, 13, and 16 (Khan et al., 2012). The most significant hotspot with the largest number of metabolites (i.e. 33 peel-related, 17 flesh-related) was on chromosome 16, which contains the *LAR1* gene, as was detected by Chagné et al. (2012b). Similarly, this hotspot co-localizes with the acidity *Ma* locus. Khan et al. (2012) summarized a list of metabolic loci, allele combinations for increased metabolite levels, as well as allelic interactions.

A phenomenon related to degradation (or oxidation) of polyphenols is enzymatic browning. Also known as flesh browning (not to be confused with internal browning), this can occur to exposed or cut apples due to oxidation of polyphenols to quinones and their subsequent polymerization (Murata et al., 1995). Browning-resistant apples (Arctic® or Arctic™ apple) were developed by Okanagan Specialty Fruits Inc. (British Columbia, Canada) through silencing of four polyphenol oxidase genes to prevent bruising during transportation and oxidation after cutting (Carter, 2012). Major loci for multiple browning-related traits (e.g. flesh browning, juice browning, polyphenol oxidase activity) were mapped on a similar region of chromosome 16, explaining up to 76.9% of phenotypic variance (Tazawa et al., 2019). A candidate gene in this region is *Leucoanthocyanidin reductase 1* (*LAR1*), a key gene in the polyphenol biosynthesis pathway. Tazawa et al. (2019) characterized non-browning haplotypes and reported a microsatellite marker, $LAR_{SSR}1\_11$ that can be used to select the non-browning allele (i.e. 284 bp). Coincidentally, *LAR1* is positioned 226 kbp downstream of the acidity *Ma* gene, and 346 kbp upstream of the *Pale Green Lethal* (*PGL*) gene (Orcheski et al., 2015; Tazawa et al., 2019). Selecting for non-browning that is tightly linked with these traits might present additional breeding challenges (e.g. seedling lethality, a non-browning apple with severe chlorophyll deficiency and/or tasting too sour).

4.3 Storability

Breeding for improved fruit storability is a major target in many apple breeding programs. While fruit is harvested once a year, providing consumers with a year-round supply of high-quality apples requires not only storage technologies but also fruits with improved storability (i.e. retention of desirable firmness after prolonged storage). Once the fruit is harvested, it undergoes a plethora of metabolic and physiochemical changes during storage, such as undesirable changes in appearance (e.g. color, shriveling), composition (e.g. water content), texture (e.g. softening, loss of crispness), flavor, and nutritional content (Watkins, 2017). Reduction in postharvest fruit quality can also arise from bruising, pathogen infection, and physiological disorders (e.g. bitter pit, superficial scald). Maintenance of optimal fruit quality after harvest is critical to growers, packers, and ultimately consumers.

Early strategies have focused on selecting apple varieties with reduced production of ethylene (Colgan et al., 2006; Zhu and Barritt, 2008), a gaseous hormone that regulates ripening and is implicated with several physiological changes of flesh texture, skin color, and flavor/aroma (Picton et al., 1995). The ethylene biosynthesis pathway consists of two large gene families coding for 1-aminocyclopropane-1-carboxylic acid synthase (ACS) and 1-aminocyclopropane-1-carboxylic acid oxidase (ACO). The precursor *S*-adenosyl-L-methionine is converted by ACS to the intermediate 1-aminocyclopropane-1-carboxylic acid (ACC), which is subsequently oxidized by ACO to produce ethylene (Yang and Hoffman, 1984). Early genomic work identified alleles associated with ethylene production in apple (Sunako et al., 1999; Harada et al., 2000) and mapped *ACS* and *ACO* genes, designated as *Malus domestica (Md)-ACS* and *Md-ACO* (Costa et al., 2005). While *ACS* and *ACO* independently affect ethylene production, *ACS* exhibited a stronger influence than *ACO* (Oraguzie et al., 2004; Costa et al., 2005). Flesh firmness is also impacted by *ACS* and *ACO*, where the presence of one favorable allele (e.g. *Md-ACS1-2* or *Md-ACO1-1*, respectively) is associated with firmer apples (Oraguzie et al., 2004; Costa et al., 2005; Zhu and Barritt, 2008). As expected, flesh firmness is also more strongly influenced by *ACS* than *ACO* (Costa et al., 2005; Zhu and Barritt, 2008). Given the dominant gene action of *ACS* and *ACO*, breeders targeting at least one favorable allele in all resultant seedlings should use a breeding parent homozygous for the favorable allele (e.g. 'Fuji' or 'Gala' for *Md-ACS1-2*). However, early maturing accessions/varieties may exhibit a higher rate of ethylene production (and therefore fruit softening) regardless of their *Md-ACS1* genotypes (Harada et al., 2000; Oraguzie et al., 2004). Allelic variations of other *ACS* genes, such as *Md-ACS3a* may also influence ethylene production and fruit softening (Wang et al., 2009; Dougherty et al., 2016). Performance predictive assays (i.e. DNA tests) are available for *Md-ACS1* (Baumgartner et al., 2016; Dougherty et al., 2016), *Md-ACO1* (Baumgartner et al., 2016), and *Md-ACS3a* (Dougherty et al., 2016), as summarized by Evans and Peace (2017).

Fruit storability and texture are also impacted by cell wall-modifying genes. During fruit ripening and softening, coordinated textural/structural changes begin with a dissolution of pectin polysaccharides of the middle lamella, followed by a disruption in the primary cell wall (Cosgrove, 2000;

Brummell, 2006). A major enzyme involved in modifying and solubilizing pectin polysaccharides is polygalacturonase (PG) (Wu et al., 1993; Hadfield and Bennett, 1998). In apple studies, *Md-PG1* has been shown to affect fruit softening during ripening (Wakasa et al., 2006; Mann et al., 2008; Costa et al., 2010; Nybom et al., 2013). Functional assessment of 14 apple cultivars showed that the rate of firmness loss was in parallel with increased expression of *Md-PG1* (Wakasa et al., 2006), although a basal level of ethylene is required to induce PG transcription (Brummell and Harpster, 2001; Costa et al., 2010). *Md-PG1* was mapped on chromosome 10 in a major QTL hotspot underpinning several fruit texture sub-phenotypes (Longhi et al., 2012). In a subsequent haplotype analysis (constructed from 11 SNPs) on 77 apple cultivars, two out of three haplotypes (i.e. H2 and H3) were linked with the absence of allele 298 from microsatellite marker Md-PG1$_{SSR}$10kd, while H1 was linked with allele 298 (Longhi et al., 2013b). Allele 298, also known as 'allele 3', was associated with loss of fruit texture in a dosage-dependent manner (Longhi et al., 2013b), and was validated for its association with flesh mealiness among Japanese apple accessions (Moriya et al., 2017b). Md-PG1$_{SSR}$10kd was touted to be a reliable marker for choosing parents without 'allele 3' and for culling seedlings with the allele (Longhi et al., 2013a). Performance predictive assays are available for *Md-PG1* (Longhi et al., 2013a; Longhi et al., 2013b; Nybom et al., 2013; Baumgartner et al., 2016).

Expansins represent another cluster of genes that is critical in depolymerizing polysaccharides during the initial stage of cell wall loosening (Cosgrove, 2000). An expansin homolog, *Md-Exp7* was mapped on chromosome 1, and a functional marker was developed based on a microsatellite motif within this gene (Costa et al., 2008). In a study of 31 apple varieties, 3 alleles were reported: 198 base pairs (bp; best firmness retention), 202 bp (intermediate), and 214 bp (worst), although post-storage firmness losses of the latter two were not statistically significant (Costa et al., 2008). Despite the 198-bp allele being associated with best firmness retention, a survey of *Md-Exp7* allelic diversity of 158 apple cultivars indicated a marked increase in allele frequency of 202 bp (from 60% to 85%) among cultivars released post-1960s compared to pre-1960s (Nybom et al., 2012).

Collectively, the interplay and epistatic interaction of five genes (i.e. *Md-ACS1*, *Md-ACS3a*, *Md-ACO1*, *Md-PG1*, and *Md-Exp7*) on storability are highly complex. Some cultivars (e.g. 'Elstar', 'Elan', and 'Rubin') exhibited excellent firmness retention despite having a double homozygous *ACS-ACO* genotype for high ethylene production, while others (e.g. 'Ben Davis' and 'Durello di Forli') the opposite (Costa et al., 2008). Nybom et al. (2013) noted that combined analyses of four markers showed low predictive power, and thus, were not cost-effective for DNA-informed breeding, unless allele frequencies of the breeding germplasm were optimized beforehand. Large-scale screenings of accessions using four markers can be difficult to analyze due to highly skewed distribution of multi-locus configurations (Nybom et al., 2013). Additionally, genetic loci for several sensory attributes (i.e. crispness, juiciness, sponginess, overall liking) were mapped to a region with the acidity locus *Ma* (Maliepaard et al., 1998; King et al., 2000). Differences in phenotyping protocols (e.g. proxy traits for storability, timing of measurements), storage regimens (e.g. duration and conditions), choice of cultivars (e.g. early versus late maturity) further complicate result comparisons across studies.

Bitter pit is a physiological disorder characterized by soft, discrete pitting (dries out and turns brown) of cortical flesh due to cell plasma membrane breakdown (Faust and Shear, 1968) (Fig. 4). Bitter pit susceptibility is associated with low calcium concentrations and poor calcium distribution within the fruit (Ferguson and Watkins, 1989), and symptoms are typically observed in fruit after storage. While the disorder incidence was shown to be variety specific, there is a strong environmental/management influence on the expression of symptoms, with varieties exhibiting a range of severities with respect to varying years and geographical locations (Ferguson and Watkins, 1989; Volz et al., 2006; Buti et al., 2015). Significant genetic variation in bitter pit expression was found among segregating families (Volz et al., 2006), suggesting that genetic improvement can be made to develop bitter pit-resistant varieties. A two-gene model for bitter pit control was postulated by Korban and Swiader (1984) but has not been confirmed in subsequent studies. Buti et al. (2015) developed a segregating family using a resistant and a mildly susceptible-to-bitter-pit parent

Figure 4 Fruit bitter pit, while typically observed after storage, can also be observed in field during fruit development.

but detected a phenotypic segregation ratio of 1:1 (resistant:susceptible) instead of a 2:1:1 (resistant:mildly susceptible:highly susceptible) ratio of a two-gene model. In addition, only one bitter pit susceptibility locus was identified by Buti et al. (2015). A major locus (*Bp-2*) for bitter pit susceptibility was mapped on chromosome 16 in two independent studies that co-localized with similar markers (Kumar et al., 2013; Buti et al., 2015). This major locus underlies a genomic hotspot for other trait loci, namely the acidity *Ma* locus (Maliepaard et al., 1998; Xu et al., 2012), fruit splitting (Kumar et al., 2013), fruit weight (Kostick et al., 2022), and to a lesser extent, internal browning, and weighted cortical intensity - an estimate of red pigment content in the fruit (Kumar et al., 2013). Performance predictive assays are available for *Bp-2*; microsatellite markers, Hi22f06 and CH05c06 with allelic combinations of 236/236 and 115/123 (respectively) are associated with high bitter pit susceptibility (Buti et al., 2015). Given the inverse association between flesh calcium concentrations and bitter pit incidence, a breeder may be tempted to use the former as an indirect selection index. However, Volz et al. (2006) cautioned against this practice due to weak correlation between these traits across families.

Soft scald, or ribbon scald, is a low-temperature storage disorder characterized by distinct darkening of skin that can extend into flesh to cause secondary infections (Hopkirk and Wills, 1981; Meheriuk et al., 1994) (Fig. 5). Various environmental/management factors have been implicated with soft scald severity, such as growing season, orchard location, storage condition, crop load, and soil fertility (Meheriuk et al., 1994; Tong et al., 2003). High narrow-sense heritability (h^2 = 0.73) of soft scald (Volz et al., 2001) and a wide range of susceptibility among varieties suggest breeding potential for varieties with reduced soft scald susceptibility. Two soft scald susceptibility

Figure 5 Soft scald or ribbon scald is a low-temperature storage disorder of apple fruit.

loci were mapped on chromosomes 2 and 3 (McClure et al., 2016). In a separate study, Howard et al. (2018) identified a major locus, described as high disorder incidence (HDI), on chromosome 2 that was likely the same chromosome 2 locus reported by McClure et al. (2016), given the proximity of loci by physical positions. Subsequently, an HDI haplotype defined by 12 SNP markers was developed using Kompetitive Allele Specific PCR (KASP™) genotyping technology (Miller et al., 2021) that can be used for DNA-informed breeding, such as choosing breeding parents or selecting/culling seedlings. Within this haplotype, two markers - SNP_FB_0452022 and SNP_FB_0949693 showed the highest accuracies in marker validation of 132 accessions (Miller et al., 2021). The same haplotype/markers can also be used to screen HDI among parents and seedlings for soggy breakdown (Miller et al., 2021), a low-temperature internal disorder characterized by fruit flesh turning brown, soft, spongy, and often moist (Plagge and Maney, 1937). Soggy breakdown is physiologically distinct from other forms of internal browning, and can occur in the absence of soft scald symptoms, and vice versa (Plagge and Maney, 1937).

A major disease affecting storability of apple fruit is blue mold (*Penicillium expansum* Link.), with an estimated annual loss in postharvest decays of over $4 million in the US apple industry (Rosenberger, 1997). While a wide range of toxic metabolites produced by the pathogen (Andersen et al., 2004) can be treated with chemical fungicides (Prusky et al., 1985), residual chemicals in fresh produce remain a concern to regulatory agencies and consumers (Cheng et al., 2016). Developing blue mold-resistant varieties can provide a sustainable and environmentally friendly solution. A locus for resistance to blue mold, named qM-*Pe*3.1, was mapped on chromosome 3 of *Malus sieversii* (Norelli et al., 2017), a wild accession with inferior fruit quality. A microsatellite-based DNA test (i.e. *Md-Pe*3) targeting qM-*Pe*3.1 was developed, where 193-bp allele was associated with smaller blue mold-induced lesion diameters (Luo et al., 2020). This test captured 16% of phenotypic variation (Luo et al., 2020), which was less than the 27.5% phenotypic variation previously reported in the same F_1 family (Norelli et al., 2017), likely due to the smaller sample size (i.e. 74 versus 187).

In addition to diagnostic markers developed/validated in trait-specific studies (covered earlier), marker applicability and reliability were also tested in pilot studies and collaborative genotyping consortia/efforts. In a pilot study of the FP7 European project FruitBreedomics (http://fruitbreedomics.com/; Laurens et al., 2010), KASP™ genotyping assays were developed for DNA-informed breeding of fruit quality traits (i.e. firmness, texture, storability), as well as resistance to apple scab, powdery mildew, and rosy apple aphid (Baumgartner et al., 2016).

The apple International RosBREED SNP Consortium OpenArray v1.0 (IRSCOA v1.0) assay was developed using 128 SNP markers linked to fruit quality and disease/pest resistance trait loci and was screened on commercial cultivars, heritage cultivars, elite selections, and segregating progenies (Chagné et al., 2019). Thirty-three SNPs were validated for use in DNA-informed breeding for scab resistance (of *Rvi2/Vh2*, *Rvi4/Vh4*, *Rvi6/Vf*), powdery mildew resistance (of *Pl2*), fire blight reduced susceptibility (*MR5/RLP1*), fruit texture, skin color, flavor intensity, as well as acidity (Chagné et al., 2019). Recommendations of predictive markers (described as Assay ID) either affirmed previous suggestions by Baumgartner et al. (2016) or provided replacements of markers with stronger statistical genotype-phenotype support (Chagné et al., 2019).

5 Genomewide selection

Genomewide selection (or genomic selection), described in the seminal paper by Meuwissen et al. (2001), is a potentially valuable breeding approach when targeting complex traits controlled by many small-effect loci. In contrast to bi-parental, pedigree-based, and association mapping procedures, which aim to identify markers with significant effects for a given trait, the goal of genomewide selection is to predict the performance (i.e. breeding values) of individuals using a high marker density that theoretically covers each linkage disequilibrium interval. Instead of identifying individuals with specific allelic combinations at trait loci, genomewide selection is used to select the best candidates in a given breeding population. A genomewide selection model is developed using phenotypic and genomewide marker data of individuals in a training population, which ideally are closely related to the population of untested individuals (i.e. test population). The model is then used to predict the performance (breeding values) of untested individuals using only their genomewide marker data, potentially reducing the number of individuals that need to be phenotyped. The long generation times and high phenotyping costs of apple make genomewide selection an attractive breeding approach.

Prediction accuracy of genomewide selection models in plants has been examined in many studies with most focusing on annual crops like barley, maize, wheat, soybean (reviewed by Krishnappa et al., 2021). Genomewide prediction accuracy is often estimated via the correlation coefficient (*r*) between the predicted breeding values and observed phenotypic values of the testing population, called predictive ability (discussed by Ould Estaghvirou et al., 2013). Several factors including marker density, trait heritability, number of QTLs that underlie the trait, size of training population, relatedness of training and testing populations, and statistical model can influence predictive ability (Daetwyler et al., 2008; Lian et al., 2014; Bernardo, 2016; Krishnappa et al., 2021; Cazenave et al., 2022). Although the statistical model (e.g. genomic-BLUP, ridge regression BLUP, Bayesian LASSO, BayesCπ) used affects predictive ability, studies have reported that predictive ability differences among models are often small (e.g. Kumar et al., 2012; Jung et al., 2022). Reported predictive abilities often vary significantly by trait and across studies for a given trait. For example, reported predictive abilities for SSC in apple ranged from less than 0.2 to ~0.9 across studies (Kumar et al., 2012; McClure et al., 2018; Minamikawa et al., 2021; Jung et al., 2022).

Several studies have estimated predictive abilities of genomewide selection models for apple appearance and eating quality traits with predictive abilities varying from study to study depending on the trait, germplasm, and/or model applied (Kumar et al., 2012, 2015; Muranty et al., 2015; Migicovsky et al., 2016; McClure et al., 2018, 2019; Roth et al., 2020; Zheng et al., 2020; Minamikawa et al., 2021; Cazenave et al., 2022; Jung et al., 2022). Examples of apple appearance and eating quality traits with predictive abilities estimated in genomewide selection studies, include percent red overcolor (less than 0.2 to 0.8; Muranty et al., 2015; Migicovsky et al., 2016; McClure et al., 2018; Minamikawa et al., 2021; Zheng et al., 2020; Cazenave et al., 2022), fruit weight (less than 0.2 to 0.8; Kumar et al., 2015; Migicovsky et al., 2016; McClure et al., 2018; Cazenave et al., 2022; Jung et al., 2022), russet formation (0.0 to 0.9; Kumar et al., 2012; Muranty et al., 2015; Migicovsky et al., 2016; Minamikawa et al., 2021; Jung et al., 2022), multiple fruit texture traits (−0.3 to 0.8; Kumar et al., 2012, 2015; Migicovsky et al., 2016; McClure et al., 2018; Roth et al., 2020; Minamikawa et al., 2021; Cazenave et al., 2022; Jung et al., 2022), SSC (less than 0.2 to 0.9; Kumar et al., 2012; Migicovsky et al., 2016; McClure et al., 2018; Minamikawa et al., 2021; Jung et al., 2022), titratable acidity (less than 0.2 to 0.8; Kumar et al., 2012; McClure et al., 2018; Jung et al., 2022), and multiple polyphenols (−0.2 to 0.5; McClure et al., 2019). Due to the wide variation in reported predictive abilities for appearance and eating quality traits in apple, predictive abilities for target traits should be validated in representative breeding germplasm before incorporating genomewide selection into an apple breeding scheme.

Similar to other crop species (discussed by Bernardo, 2016), the primary focus of most genomewide prediction studies in apple has been the estimation of model accuracy in apple germplasm, not implementation as a breeding strategy in breeding programs. Although genomewide selection is a promising breeding strategy, adoption has been slow in rosaceous crop breeding programs. Strawberry (*Fragaria × ananassa*) breeding is an exception where genomewide selection has successfully been implemented into breeding schemes (e.g. Osorio et al., 2021; Whitaker et al., 2017). For genomewide selection to become routine, breeding programs need: (1) cost-effective, efficient, robust genotyping platforms and providers; (2) user-friendly data curation and analysis software and/or trained personnel; and (3) an efficient workflow for integration of genomewide predictions into the breeding scheme (described by Bernardo, 2016). Multiple SNP arrays (Chagné et al., 2012a; Bianco et al., 2014, 2016) are available for apple and are routinely used to genotype advanced selections and/or mapping families in apple breeding programs. Wide variation in reported trait predictive abilities and relatively high cost per genotyping sample might continue to limit adoption of genomewide selection in apple breeding programs.

6 Conclusion and future trends

This chapter focuses on DNA-informed breeding as genetic and genomic tools to increase the efficiency and accuracy of breeding for fruit quality improvement in an apple breeding program. Further acceleration in breeding efficiency can be achieved by shortening the apple's juvenility period (i.e. 5–10 years). A strategy for accelerated trait introgression is rapid-cycle or fast-track breeding, which was comprehensively reviewed by Callahan et al. (2016). The most common approach utilizes a transgenic intermediate generation that overexpresses an early/precocious flowering gene (e.g. *BpMADS4* from silver birch or *Betula pendula* Roth.) to shorten the juvenility period to 3–8 months, enabling the development of one generation cycle per year (Flachowsky et al., 2007, 2011). This approach can speed up pyramiding of multiple resistance genes into an elite genetic background. To date, rapid-cycle breeding has targeted disease resistance traits (i.e. allele pyramiding through successive crosses) with the eventual goal of crossing the pyramided transgenic line with an elite breeding parent (Khan and Korban, 2022). Flachowsky et al. (2011) utilized a *Bp-MADS4*-transgenic apple line, T1190 to expedite introgression of apple scab *Rvi2*, *Rvi4* (both from donor 'Regia'), powdery mildew *Pl-1*, *Pl-2* (both from accession 98/6-10) resistance genes, and the fire blight reduced susceptibility locus FB-F7 (from 'Regia'). In a separate work, T1190 was used to accelerate marker-assisted introgression of the fire blight reduced susceptibility locus Fb_E from 'Evereste' (Le Roux et al., 2012). Non-transgenic null segregants of the fifth generation (i.e. modified backcross 4) carrying the Fb_E locus was developed within 7 years, while retaining only 4% of the donor's (i.e. 'Evereste') genome (Schlathölter et al., 2018). Successful implementation of rapid-cycle breeding requires diagnostic markers for traits of interest (e.g. disease resistance, pest resistance, fruit quality). In later cycles, selections/seedlings will be selected for the trait loci/genes of interest, but against the early-flowering gene, which may require genome sequencing at high depths (e.g. 100×) to detect presence of any transgenic content (Khan and Korban, 2022). Depending on future and evolving regulatory frameworks, selections lacking the *BpMADS4* gene may be deemed non-transgenic and/or deregulated for commercial release (USDA, 2011; Schlathölter et al., 2018).

7 Where to look for further information

General information on apple breeding was described by Hancock et al. (2008), Iwanami (2011), and Brown (2012). An overview to apple genomics was covered by Korban and Tartarini (2009), and Dhingra (2017). Teh et al. (2021b) provided a summary of key apple genetic loci. A primer to DNA-informed breeding and applications was described by Evans and Peace (2017). Other chapters of the book *Achieving Sustainable Cultivation of Apples* (Evans, 2017) that may be of interest include disease/pest resistance-based breeding approaches, rootstock breeding, horticultural management, and postharvest fruit management. Current biotechnology tools that are relevant to apple breeding were described by Hanke et al. (2020). Concepts, advances, and applications of rapid-cycle breeding were detailed by Callahan et al. (2016).

The European Association for Research on Plant Breeding (EUCARPIA; https://www.eucarpia.eu/) Fruit section meets every four years and publishes proceedings as Acta Horticulturae (https://www.actahort.org/). The Rosaceae International Genomics Initiative research community meets every two years at the International Rosaceae Genomics Conference. Information about this group and other Rosaceae community conferences can be found on the Genome Database for Rosaceae (GDR) website (https://www.rosaceae.org/; Jung et al., 2019). In addition, this community website houses a searchable genome database, publication databases, various genomic tools, as well as links to other community projects, such as RosBREED (https://www.rosbreed.org/; Iezzoni et al., 2010). Relevant information can also be found through annual national meetings/conferences of societies, namely the American Society of Horticultural Science (ASHS; https://ashs.org/), and the American Pomological Society (APS; https://www.americanpomological.org/).

8 References

Amyotte, B., Bowen, A. J., Banks, T., Rajcan, I. and Somers, D. J. (2017). Mapping the sensory perception of apple using descriptive sensory evaluation in a genome wide association study, *PLoS ONE* 12(2):e0171710.

Andersen, B., Smedsgaard, J. and Frisvad, J. C. (2004). *Penicillium expansum*: Consistent production of patulin, chaetoglobosins, and other secondary metabolites in culture and their natural occurrence in fruit products, *J. Agric. Food Chem.* 52(8):2421–2428.

Arts, I. C. and Hollman, P. C. (2005). Polyphenols and disease risk in epidemiologic studies, *Am. J. Clin. Nutr.* 81(1):317S–325S.

Bai, Y., Dougherty, L., Li, M., Fazio, G., Cheng, L. and Xu, K. (2012). A natural mutation-led truncation in one of the two aluminum-activated malate transporter-like genes at the *Ma* locus is associated with low fruit acidity in apple, *Mol. Genet. Genomics* 287(8):663–678.

Ban, Y., Honda, C., Hatsuyama, Y., Igarashi, M., Bessho, H. and Moriguchi, T. (2007). Isolation and functional analysis of a MYB transcription factor gene that is a key regulator for the development of red coloration in apple skin, *Plant Cell Physiol.* 48(7):958-970.

Baumgartner, I. O., Kellerhals, M., Costa, F., Dondini, L., Pagliarani, G., Gregori, R., Tartarini, S., Leumann, L., Laurens, F. and Patocchi, A. (2016). Development of SNP-based assays for disease resistance and fruit quality traits in apple (*Malus × domestica* Borkh.) and validation in breeding pilot studies, *Tree Genet. Genomes* 12:1-21.

Bernardo, R. (2016). Bandwagons I, too, have known, *Theor. Appl. Genet.* 129(12):2323-2332.

Bianco, L., Cestaro, A., Linsmith, G., Muranty, H., Denancé, C., Théron, A., Poncet, C., Micheletti, D., Kerschbamer, E., Di Pierro, E. A., Larger, S., Pindo, M., Van de Weg, E., Davassi, A., Laurens, F., Velasco, R., Durel, C. E. and Troggio, M. (2016). Development and validation of the Axiom® Apple480K SNP genotyping array, *Plant J.* 86(1):62-74.

Bianco, L., Cestaro, A., Sargent, D. J., Banchi, E., Derdak, S., Di Guardo, M., Salvi, S., Jansen, J., Viola, R., Gut, I., Laurens, F., Chagné, D., Velasco, R., van de Weg, E. and Troggio, M. (2014). Development and validation of a 20K single nucleotide polymorphism (SNP) whole genome genotyping array for apple (*Malus × domestica* Borkh), *PLoS ONE* 9(10):e110377.

Bliss, F. A. (2010). Marker-assisted breeding in horticultural crops, *Acta Hortic.* 859:339-350.

Brown, S. (2012). Apple. In M. L. Badenes and D. H. Byrne (Eds.), *Fruit Breeding*. New York: Springer, pp. 329-367.

Brummell, D. A. (2006). Cell wall disassembly in ripening fruit, *Funct. Plant Biol.* 33(2):103-119.

Brummell, D. A. and Harpster, M. H. (2001). Cell wall metabolism in fruit softening and quality and its manipulation in transgenic plants, *Plant Mol. Biol.* 47(1-2):311-340.

Buti, M., Poles, L., Caset, D., Magnago, P., Fernández-Fernández, F., Colgan, R. J., Velasco, R. and Sargent, D. J. (2015). Identification and validation of a QTL influencing bitter pit symptoms in apple (*Malus × domestica*), *Mol. Breed.* 35(1):29.

Callahan, A. M., Srinivasan, C., Dardick, C., Scorza, R., Goldman, I. L. and Ortiz, R. (2016). Rapid cycle breeding: Application of transgenic early flowering for perennial trees, *Plant Breed. Rev.* 40:299-334.

Cappellin, L., Farneti, B., Di Guardo, M., Busatto, N., Khomenko, I., Romano, A., Velasco, R., Costa, G., Biasioli, F. and Costa, F. (2015). QTL analysis coupled with PTR-ToF-MS and candidate gene-based association mapping validate the role of Md-AAT1 as a major gene in the control of flavor in apple fruit, *Plant Mol. Biol. Rep.* 33(2):239-252.

Carter, N. (2012). *Petition for Determination of Nonregulated Status: Arctic™ Apple (Malus × domestica) Events GD743 and GS784.* United States Department of Agriculture - Animal and Plant Health Inspection Service.

Cazenave, X., Petit, B., Lateur, M., Nybom, H., Sedlak, J., Tartarini, S., Laurens, F., Durel, C. E. and Muranty, H. (2022). Combining genetic resources and elite material populations to improve the accuracy of genomic prediction in apple, *Genet* 12:jkab420.

Chagné, D., Carlisle, C. M., Blond, C., Volz, R. K., Whitworth, C. J., Oraguzie, N. C., Crowhurst, R. N., Allan, A. C., Espley, R. V., Hellens, R. P. and Gardiner, S. E. (2007). Mapping a candidate gene (*MdMYB10*) for red flesh and foliage colour in apple, *BMC Genomics* 8:212.

Chagné, D., Crowhurst, R. N., Troggio, M., Davey, M. W., Gilmore, B., Lawley, C., Vanderzande, S., Hellens, R. P., Kumar, S., Cestaro, A., Velasco, R., Main, D., Rees, J. D., Iezzoni, A., Mockler, T., Wilhelm, L., Van de Weg, E., Gardiner, S. E., Bassil, N. and Peace, C. (2012a). Genome-wide SNP detection, validation, and development of an 8K SNP array for apple, *PLoS ONE* 7(2):e31745.

Chagné, D., Krieger, C., Rassam, M., Sullivan, M., Fraser, J., André, C., Pindo, M., Troggio, M., Gardiner, S. E., Henry, R. A., Allan, A. C., McGhie, T. K. and Laing, W. A. (2012b). QTL and candidate gene mapping for polyphenolic composition in apple fruit, *BMC Plant Biol.* 12:12.

Chagné, D., Kirk, C., How, N., Whitworth, C., Fontic, C., Reig, G., Sawyer, G., Rouse, S., Poles, L., Gardiner, S. E., Kumar, S., Espley, R., Volz, R. K., Troggio, M. and Iglesias, I. (2016). A functional genetic marker for apple red skin coloration across different environments, *Tree Genet. Genomes* 12(4):67.

Chagné, D., Vanderzande, S., Kirk, C., Profitt, N., Weskett, R., Gardiner, S. E., Peace, C. P., Volz, R. K. and Bassil, N. V. (2019). Validation of SNP markers for fruit quality and disease resistance loci in apple (*Malus × domestica* Borkh.) using the OpenArray® platform, *Hortic. Res.* 6:30.

Cheng, F. S., Weeden, N. F. and Brown, S. K. (1996). Identification of co-dominant RAPD markers tightly linked to fruit skin color in apple, *Theor. Appl. Genet.* 93(1-2):222-227.

Cheng, L., Jiang, S., Zhang, S., You, H., Zhang, J., Zhou, Z., Xiao, Y., Liu, X., Du, Y., Li, J., Wang, X., Xin, Y., Zheng, Y. and Shang, K. (2016). Consumers' behaviors and concerns on fresh vegetable purchase and safety in Beijing urban areas, China, *Food Control* 63:101-109.

Colgan, R., Stow, J., Marchese, A., Nidzovic, S. and Else, M. (2006). Storage quality of low ethylene producing apples, *J. Fruit Ornam Plant Res.* 14:85-92.

Collard, B. C. Y. and Mackill, D. J. (2008). Marker-assisted selection: An approach for precision plant breeding in the twenty-first century, *Phil. Trans. R. Soc. B Biol. Sci.* 363(1491):557-572.

Corelli-Grappadelli, L. and Lakso, A. N. (2004). Fruit development in deciduous tree crops as affected by physiological factors and environmental conditions (KEYNOTE), *Acta Hortic.* 636:425-441.

Cosgrove, D. J. (2000). Loosening of plant cell walls by expansins, *Nature* 407(6802):321-326.

Costa, F., Cappellin, L., Longhi, S., Guerra, W., Magnago, P., Porro, D., Soukoulis, C., Salvi, S., Velasco, R., Biasioli, F. and Gasperi, F. (2011). Assessment of apple (*Malus × domestica* Borkh.) fruit texture by a combined acoustic-mechanical profiling strategy, *Postharvest Biol. Technol.* 61(1):21-28.

Costa, F., Peace, C. P., Stella, S., Serra, S., Musacchi, S., Bazzani, M., Sansavini, S. and Van de Weg, W. E. (2010). QTL dynamics for fruit firmness and softening around an ethylene-dependent polygalacturonase gene in apple (*Malus × domestica* Borkh.), *J. Exp. Bot.* 61(11):3029-3039.

Costa, F., Stella, S., van de Weg, W. E., Guerra, W., Cecchinel, M., Dallavia, J., Koller, B. and Sansavini, S. (2005). Role of the genes *Md-ACO1* and *Md-ACS1* in ethylene production and shelf life of apple (*Malus domestica* Borkh), *Euphytica* 141(1-2):181-190.

Costa, F., van de Weg, W. E., Stella, S., Dondini, L., Pratesi, D., Musacchi, S. and Sansavini, S. (2008). Map position and functional allelic diversity of *Md-Exp7*, a new putative expansin gene associated with fruit softening in apple (*Malus × domestica* Borkh.) and pear (*Pyrus communis*), *Tree Genet. Genomes* 4(3):575-586.

Daccord, N., Celton, J. M., Linsmith, G., Becker, C., Choisne, N., Schijlen, E., van de Geest, H., Bianco, L., Micheletti, D., Velasco, R., Di Pierro, E. A., Gouzy, J., Rees, D. J. G., Guérif, P., Muranty, H., Durel, C. E., Laurens, F., Lespinasse, Y., Gaillard, S., Aubourg, S., Quesneville, H., Weigel, D., van de Weg, E., Troggio, M. and Bucher, E. (2017). High-quality de novo assembly of the apple genome and methylome dynamics of early fruit development, *Nat. Genet.* 49(7):1099–1106.

Daetwyler, H. D., Villanueva, B. and Woolliams, J. A. (2008). Accuracy of predicting the genetic risk of disease using a genome-wide approach, *PLoS ONE* 3(10):e3395.

Dhingra, A. (2017). The apple genome – Harbinger of innovation for sustainable apple production. In K. Evans (Ed.), *Achieving Sustainable Cultivation of Apples*. Pennsylvania: Burleigh Dodds Science Publishing, pp. 23–34.

Di Guardo, M., Bink, M. C. A. M., Guerra, W., Letschka, T., Lozano, L., Busatto, N., Poles, L., Tadiello, A., Bianco, L., Visser, R. G. F., van de Weg, E. and Costa, F. (2017). Deciphering the genetic control of fruit texture in apple by multiple family-based analysis and genome-wide association, *J. Exp. Bot.* 68(7):1451–1466.

Dixon, J. and Hewett, E. W. (2000). Factors affecting apple aroma/flavour volatile concentration: A review, *New Zeal J. Crop Hort.* 28(3):155–173.

Dougherty, L., Zhu, Y. and Xu, K. (2016). Assessing the allelotypic effect of two aminocyclopropane carboxylic acid synthase-encoding genes *MdACS1* and *MdACS3a* on fruit ethylene production and softening in *Malus*, *Hortic. Res.* 3:16024.

Duan, N., Bai, Y., Sun, H., Wang, N., Ma, Y., Li, M., Wang, X., Jiao, C., Legall, N., Mao, L., Wan, S., Wang, K., He, T., Feng, S., Zhang, Z., Mao, Z., Shen, X., Chen, X., Jiang, Y., Wu, S., Yin, C., Ge, S., Yang, L., Jiang, S., Xu, H., Liu, J., Wang, D., Qu, C., Wang, Y., Zuo, W., Xiang, L., Liu, C., Zhang, D., Gao, Y., Xu, Y., Xu, K., Chao, T., Fazio, G., Shu, H., Zhong, G. Y., Cheng, L., Fei, Z. and Chen, X. (2017). Genome re-sequencing reveals the history of apple and supports a two-stage model for fruit enlargement, *Nat. Commun.* 8(1):249.

Dunemann, F., Ulrich, D., Boudichevskaia, A., Grafe, C. and Weber, W. E. (2009). QTL mapping of aroma compounds analysed by headspace solid-phase microextraction gas chromatography in the apple progeny 'Discovery' × 'Prima', *Mol. Breed.* 23:501–521.

Dunemann, F., Ulrich, D., Malysheva-Otto, L., Weber, W. E., Longhi, S., Velasco, R. and Costa, F. (2012). Functional allelic diversity of the apple alcohol acyl-transferase gene *MdAAT1* associated with fruit ester volatile contents in apple cultivars, *Mol. Breed.* 29(3):609–625.

Edge-Garza, D. A., Luby, J. J. and Peace, C. (2015). Decision support for cost-efficient and logistically feasible marker-assisted seedling selection in fruit breeding, *Mol. Breed.* 35(12):223.

Edge-Garza, D. A., Peace, C. P. and Yanmin Zhu (2010). Enabling marker-assisted seedling selection in the Washington apple breeding program, *Acta Hortic.* 859:369–373.

El Hadi, M. A. M., Zhang, F. J., Wu, F. F., Zhou, C. H. and Tao, J. (2013). Advances in fruit aroma volatile research, *Molecules* 18(7):8200–8229.

Espley, R. V., Hellens, R. P., Putterill, J., Stevenson, D. E., Kutty-Amma, S. and Allan, A. C. (2007). Red colouration in apple fruit is due to the activity of the MYB transcription factor, MdMYB10, *Plant J.* 49(3):414–427.

Evans, K. (2013). Apple breeding in the Pacific Northwest, *Acta Hortic.* 976:75–78.

Evans, K. (2017). *Achieving Sustainable Cultivation of Apples*. Pennsylvania: Burleigh Dodds Science Publishing.

Evans, K., Brutcher, L., Konishi, B. and Barritt, B. (2010). Correlation of sensory analysis with physical textural data from a computerized penetrometer in the Washington State University apple breeding program, *hortTechnology* 20(6):1026–1029.

Evans, K., Konishi, B., Brutcher, L., Edge-Garza, D., Rowland, T. and Peace, C. (2012). The logistical challenges of marker-assisted seedling selection in an apple breeding program. In *6th Rosaceous Genomics Conference*, Italy, pp. 92–93.

Evans, K. and Peace, C. (2017). Advances in marker-assisted breeding of apples. In K. Evans (Ed.), *Achieving Sustainable Cultivation of Apples*. Pennsylvania: Burleigh Dodds Science Publishing, pp. 165–192.

Faust, M. and Shear, C. B. (1968). Corking disorders of apples: A physiological and biochemical review, *Bot. Rev.* 34(4):441–469.

Ferguson, I. B. and Watkins, C. B. (1989). Bitter pit in apple fruit, *Hortic. Rev.* 11:289–355.

Flachowsky, H., Le Roux, P. M., Peil, A., Patocchi, A., Richter, K. and Hanke, M. V. (2011). Application of a high-speed breeding technology to apple (*Malus* × *domestica*) based on transgenic early flowering plants and marker-assisted selection, *New Phytol.* 192(2):364–377.

Flachowsky, H., Peil, A., Sopanen, T., Elo, A. and Hanke, V. (2007). Overexpression of *BpMADS4* from silver birch (*Betula pendula* Roth.) induces early-flowering in apple (*Malus* × *domestica* Borkh.), *Plant Breed.* 126(2):137–145.

FAOSTAT. (2020). Crops and livestock products. *Food and Agriculture Organization of the United Nations Statistical Database*. Available at: https://www.fao.org/faostat/en/#data/QCL/ (accessed 15 February 2022).

Fuleki, T., Pelayo, E. and Palabay, R. B. (1994). Sugar composition of varietal juices produced from fresh and stored apples, *J. Agric. Food Chem.* 42(6):1266–1275.

Gardner, K. M., Brown, P., Cooke, T. F., Cann, S., Costa, F., Bustamante, C., Velasco, R., Troggio, M. and Myles, S. (2014). Fast and cost-effective genetic mapping in apple using next-generation sequencing, *Genet.* 4:1681–1687.

Gianfranceschi, L. and Soglio, V. (2004). The European project HiDRAS: Innovative multidisciplinary approaches to breeding high-quality disease resistant apples, *Acta Hortic.* 663:327–330.

Guan, Y., Peace, C., Rudell, D., Verma, S. and Evans, K. (2015). QTLs detected for individual sugars and soluble solids content in apple, *Mol. Breed.* 35(6):135.

Hadfield, K. A. and Bennett, A. B. (1998). Polygalacturonases: Many genes in search of a function, *Plant Physiol.* 117(2):337–343.

Hancock, J. F., Luby, J. J., Brown, S. K. and Lobos, G. A. (2008). Apples. In J. F. Hancock (Ed.), *Temperate Fruit Crop Breeding: Germplasm to Genomics*. Dordrecht, Netherlands: Springer, pp. 1–37.

Hanke, M. V., Flachowsky, H., Peil, A. and Emeriewen, O. F. (2020). 19.3 Malus × domestica apple. In R. E. Litz, F. Pliego-Alfaro and J. I. Hormaza (Eds.), *Biotechnology of Fruit and Nut Crops*. Oxfordshire, United Kingdom: CAB International, pp. 440–473.

Harada, T., Sunako, T., Wakasa, Y., Soejima, J., Satoh, T. and Niizeki, M. (2000). An allele of the 1-aminocyclopropane-1-carboxylate synthase gene (*Md-ACS1*) accounts for the low level of ethylene production in climacteric fruits of some apple cultivars, *Theor. Appl. Genet.* 101(5–6):742–746.

Harker, F. R., Kupferman, E. M., Marin, A. B., Gunson, F. A. and Triggs, C. M. (2008). Eating quality standards for apples based on consumer preferences, *Postharvest Biol. Technol.* 50(1):70–78.

Hopkirk, G. and Wills, R. B. H. (1981). Variation in fatty acid composition of apples in relation to soft scald, *Phytochemistry* 20(2):193–195.

Howard, N. P., van de Weg, E., Tillman, J., Tong, C. B. S., Silverstein, K. A. T. and Luby, J. J. (2018). Two QTL characterized for soft scald and soggy breakdown in apple (*Malus × domestica*) through pedigree-based analysis of a large population of interconnected families, *Tree Genet. Genomes* 14(1):2.

Iezzoni, A., Peace, C., Main, D., Bassil, N., Coe, M., Finn, C., Gasic, K., Luby, J., Hokanson, S., McFerson, J., Norelli, J., Olmstead, M., Whitaker, V. and Yue, C. (2017). RosBREED2: Progress and future plans to enable DNA-informed breeding in the Rosaceae, *Acta Hortic.* (1172):115–118.

Iezzoni, A., Weebadde, C., Luby, J., Chengyan Yue, van de Weg, E., Fazio, G., Main, D., Peace, C. P., Bassil, N. V. and McFerson, J. (2010). RosBREED: Enabling marker-assisted breeding in Rosaceae, *Acta Hortic.* 859:389–394.

Iwanami, H. (2011). Breeding for fruit quality in apple. In M. A. Jenks and P. J. Bebeli (Eds.), *Breeding for Fruit Quality*. New York: John Wiley & Sons, pp. 175–200.

Jia, D., Shen, F., Wang, Y., Wu, T., Xu, X., Zhang, X. and Han, Z. (2018). Apple fruit acidity is genetically diversified by natural variations in three hierarchical epistatic genes: *MdSAUR37*, *MdPP2CH* and *MdALMTII*, *Plant J.* 95(3):427–443.

Jung, M., Keller, B., Roth, M., Aranzana, M. J., Auwerkerken, A., Guerra, W., Al-Rifaï, M., Lewandowski, M., Sanin, N., Rymenants, M., Didelot, F., Dujak, C., Forcada, C. F., Knauf, A., Laurens, F., Studer, B., Muranty, H. and Patocch, A. (2022). Genetic architecture and genomic predictive ability of apple quantitative traits across environments, *Hortic. Res.* 9:uhac028.

Jung, S., Lee, T., Cheng, C. H., Buble, K., Zheng, P., Yu, J., Humann, J., Ficklin, S. P., Gasic, K., Scott, K., Frank, M., Ru, S., Hough, H., Evans, K., Peace, C., Olmstead, M., DeVetter, L. W., McFerson, J., Coe, M., Wegrzyn, J. L., Staton, M. E., Abbott, A. G. and Main, D. (2019). 15 years of GDR: New data and functionality in the Genome Database for Rosaceae, *Nucleic Acids Res.* 47(D1):D1137–D1145.

Kellerhals, M., Spuhler, M., Duffy, B., Patocchi, A. and Frey, J. E. (2009). Selection efficiency in apple breeding, *Acta Hortic.* 814:177–184.

Kenis, K., Keulemans, J. and Davey, M. W. (2008). Identification and stability of QTLs for fruit quality traits in apple, *Tree Genet. Genomes* 4(4):647–661.

Khan, A. and Korban, S. S. (2022). Breeding and genetics of disease resistance in temperate fruit trees: Challenges and new opportunities, *Theor. Appl. Genet.* 135(11):3961–3985. DOI: 10.1007/s00122-022-04093-0.

Khan, S. A., Chibon, P. Y., de Vos, R. C. H., Schipper, B. A., Walraven, E., Beekwilder, J., van Dijk, T., Finkers, R., Visser, R. G., van de Weg, E. W., Bovy, A., Cestaro, A., Velasco, R., Jacobsen, E. and Schouten, H. J. (2012). Genetic analysis of metabolites in apple fruits indicates an mQTL hotspot for phenolic compounds on linkage group 16, *J. Exp. Bot.* 63(8):2895–2908.

King, G. J., Maliepaard, C., Lynn, J. R., Alston, F. H., Durel, C. E., Evans, K. M., Griffon, B., Laurens, F., Manganaris, A. G., Schrevens, E., Tartarini, S. and Verhaegh, J. (2000). Quantitative genetic analysis and comparison of physical and sensory descriptors relating to fruit flesh firmness in apple (*Malus pumila* Mill.), *Theor. Appl. Genet.* 100(7):1074–1084.

King, M. C. and Cliff, M. A. (2002). Development of a model for prediction of consumer liking from visual attributes of new and established apple cultivars, *J. Amer. Pomolog. Soc.* 56:223–229.

Korban, S. S. and Swiader, J. M. (1984). Genetic and nutritional status in bitter pit-resistant and pit-susceptible apple seedlings, *J. Am. Soc. Hortic. Sci.* 109:428–432.

Korban, S. S. and Tartarini, S. (2009). Apple structural genomics. In K. M. Folta and S. E. Gardiner (Eds.), *Genetics and Genomics of Rosaceae*. New York: Springer, pp. 85–119.

Kostick, S. A. and Luby, J. J. (2022). Apple fruit size QTLs on chromosomes 8 and 16 characterized in 'Honeycrisp'-derived germplasm, *Agronomy* 12(6):1279.

Krishnappa, G., Savadi, S., Tyagi, B. S., Singh, S. K., Mamrutha, H. M., Kumar, S., Mishra, C. N., Khan, H., Gangadhara, K., Uday, G., Singh, G. and Singh, G. P. (2021). Integrated genomic selection for rapid improvement of crops, *Genomics* 113(3):1070–1086.

Kumar, S., Chagné, D., Bink, M. C. A. M., Volz, R. K., Whitworth, C. and Carlisle, C. (2012). Genomic selection for fruit quality traits in apple (*Malus × domestica* Borkh.), *PLoS ONE* 7(5):e36674.

Kumar, S., Garrick, D. J., Bink, M. C., Whitworth, C., Chagné, D. and Volz, R. K. (2013). Novel genomic approaches unravel genetic architecture of complex traits in apple, *BMC Genomics* 14:393.

Kumar, S., Molloy, C., Muñoz, P., Daetwyler, H., Chagné, D. and Volz, R. (2015). Genome-enabled estimates of additive and nonadditive genetic variances and prediction of apple phenotypes across environments, *Genetics* 5:2711–2718.

Larsen, B., Migicovsky, Z., Jeppesen, A. A., Gardner, K. M., Toldam-Andersen, T. B., Myles, S., Ørgaard, M., Petersen, M. A. and Pedersen, C. (2019). Genome-wide association studies in apple reveal loci for aroma volatiles, sugar composition, and harvest date, *Plant Genome* 12(2):2.

Laurens, F., Aranzana, M. J., Arús, P., Bonany, J., Corelli, L., Patocchi, A., Peil, A., Quilot, B., Salvi, S., van de Weg, E. and Vecchietti, A. (2010). FruitBreedomics: A new European initiative to bridge the gap between scientific research and breeding Rosaceae fruit tree crops. Book of Abstracts v2, International Horticulture Congress, Lisbon, p. 242.

Le Roux, P. M., Flachowsky, H., Hanke, M. V., Gessler, C. and Patocchi, A. (2012). Use of a transgenic early flowering approach in apple (*Malus × domestica* Borkh.) to introgress fire blight resistance from cultivar Evereste, *Mol. Breed.* 30(2):857–874.

Lian, L., Jacobson, A., Zhong, S. and Bernardo, R. (2014). Genomewide prediction accuracy within 969 maize biparental populations, *Crop Sci.* 54(4):1514–1522.

Longhi, S., Cappellin, L., Guerra, W. and Costa, F. (2013a). Validation of a functional molecular marker suitable for marker-assisted breeding for fruit texture in apple (*Malus × domestica* Borkh.), *Mol. Breed.* 32(4):841–852.

Longhi, S., Hamblin, M. T., Trainotti, L., Peace, C. P., Velasco, R. and Costa, F. (2013b). A candidate gene based approach validates *Md-PG1* as the main responsible for a QTL impacting fruit texture in apple (*Malus × domestica* Borkh), *BMC Plant Biol.* 13:1–13.

Longhi, S., Moretto, M., Viola, R., Velasco, R. and Costa, F. (2012). Comprehensive QTL mapping survey dissects the complex fruit texture physiology in apple (*Malus × domestica* Borkh.), *J. Exp. Bot.* 63(3):1107–1121.

Luby, J. J. and Shaw, D. V. (2001). Does marker-assisted selection make dollars and sense in a fruit breeding program?, *Hortscience* 36(5):872–879.

Luo, F., Norelli, J. L., Howard, N. P., Wisniewski, M., Flachowsky, H., Hanke, M. V. and Peace, C. (2020). Introgressing blue mold resistance into elite apple germplasm by rapid cycle breeding and foreground and background DNA-informed selection, *Tree Genet. Genomes* 16(2):28.

Maliepaard, C., Alston, F. H., Van Arkel, G., Brown, L. M., Chevreau, E., Dunemann, F., Evans, K. M., Gardiner, S., Guilford, P., van Heusden, A. W., Janse, J., Laurens, F., Lynn, J. R., Manganaris, A. G., den Nijs, A. P. M., Periam, N., Rikkerink, E., Roche, P., Ryder, C., Sansavini, S., Schmidt, H., Tartarini, S., Verhaegh, J. J., Vrielink-van Ginkel, M. and King, G. J. (1998). Aligning male and female linkage maps of apple (*Malus pumila* Mill.) using multi-allelic markers, *Theor. Appl. Genet.* 97(1–2):60–73.

Mann, H. S., Alton, J. J., Kim, S. and Tong, C. B. S. (2008). Differential expression of cell-wall–modifying genes and novel cDNAs in apple fruit during storage, *J. Am. Soc. Hortic. Sci.* 133(1):152–157.

McClure, K. A., Gardner, K. M., Douglas, G. M., Song, J., Forney, C. F., DeLong, J., Fan, L., Du, L., Toivonen, P. M. A., Somers, D. J., Rajcan, I. and Myles, S. (2018). A genome-wide association study of apple quality and scab resistance, *Plant Genom.* 11(1):170075.

McClure, K. A., Gardner, K. M., Toivonen, P. M., Hampson, C. R., Song, J., Forney, C. F., DeLong, J., Rajcan, I. and Myles, S. (2016). QTL analysis of soft scald in two apple populations, *Hortic. Res.* 3:16043.

McClure, K. A., Gong, Y., Song, J., Vinqvist-Tymchuk, M., Campbell Palmer, L., Fan, L., Burgher-MacLellan, K., Zhang, Z., Celton, J. M., Forney, C. F., Migicovsky, Z. and Myles, S. (2019). Genome-wide association studies in apple reveal loci of large effect controlling apple polyphenols, *Hortic. Res.* 6:107.

Meheriuk, M. P., Prange, R. K., Lidster, R. K. and Porritt, S. W. (1994). *Postharvest Disorders of Apples and Pears.* Ottawa, Canada: Agriculture Canada Publication 1737/E, Agriculture and Agri-Food Canada.

Meuwissen, T. H. E., Hayes, B. J. and Goddard, M. E. (2001). Prediction of total genetic value using genome-wide dense marker maps, *Genetics* 157(4):1819–1829.

Migicovsky, Z., Gardner, K. M., Money, D., Sawler, J., Bloom, J. S., Moffett, P., Chao, C. T., Schwaninger, H., Fazio, G., Zhong, G. Y. and Myles, S. (2016). Genome to phenome mapping in apple using historical data, *Plant Genom.* 9(2):2.

Miller, B. A., Tillman, J. R., Howard, N. P., Kostick, S. A., Evans, K. M. and Luby, J. J. (2021). A DNA test for high incidence of soft scald and soggy breakdown postharvest disorders in *Malus domestica* Borkh, *Mol. Breed.* 41(9):57.

Minamikawa, M. F., Kunihisa, M., Noshita, K., Moriya, S., Abe, K., Hayashi, T., Katayose, Y., Matsumoto, T., Nishitani, C., Terakami, S., Yamamoto, T. and Iwata, H. (2021). Tracing founder haplotypes of Japanese apple varieties: Application in genomic prediction and genome-wide association study, *Hortic. Res.* 8(1):49.

Moriya, S., Kunihisa, M., Okada, K., Shimizu, T., Honda, C., Yamamoto, T., Muranty, H., Denancé, C., Katayose, Y., Iwata, H. and Abe, K. (2017a). Allelic composition of *MdMYB1* drives red skin color intensity in apple (*Malus* × *domestica* Borkh.) and its application to breeding, *Euphytica* 213(4):78.

Moriya, S., Kunihisa, M., Okada, K., Iwanami, H., Iwata, H., Minamikawa, M., Katayose, Y., Matsumoto, T., Mori, S., Sasaki, H. and Matsumoto, T. (2017b). Identification of QTLs for flesh mealiness in apple (*Malus* × *domestica* Borkh.), *Horticult J.* 86:159–170.

Muranty, H., Troggio, M., Sadok, I. B., Rifaï, M. A., Auwerkerken, A., Banchi, E., Velasco, R., Stevanato, P., van de Weg, W. E., Di Guardo, M., Kumar, S., Laurens, F. and Bink, M. C. (2015). Accuracy and responses of genomic selection on key traits in apple breeding, *Hortic. Res.* 2:15060.

Murata, M., Noda, I. and Homma, S. (1995). Enzymatic browning of apples on the market: Relationship between browning, polyphenol content, and polyphenol oxidase, *J. Jpn. Soc. Food Sci. Technol.* 42(10):820–826.

Musacchi, S. and Serra, S. (2018). Apple fruit quality: Overview on pre-harvest factors, *Sci. Hortic.* 234:409–430.

Nicolai, B. M., Beullens, K., Bobelyn, E., Peirs, A., Saeys, W., Theron, K. I. and Lammertyn, J. (2007). Nondestructive measurement of fruit and vegetable quality by means of NIR spectroscopy: A review, *Postharvest Biol. Technol.* 46(2):99–118.

Norelli, J. L., Wisniewski, M., Fazio, G., Burchard, E., Gutierrez, B., Levin, E. and Droby, S. (2017). Genotyping-by-sequencing markers facilitate the identification of quantitative trait loci controlling resistance to *Penicillium expansum* in *Malus sieversii*, *PLoS ONE* 12(3):e0172949.

Nybom, H., Ahmadi-Afzadi, M., Garkava-Gustavsson, L., Sehic, J. and Tahir, I. (2012). Selection for improved fruit texture and storability in apple, *Acta Hortic.* 934:849–854.

Nybom, H., Ahmadi-Afzadi, M., Sehic, J. and Hertog, M. (2013). DNA marker-assisted evaluation of fruit firmness at harvest and post-harvest fruit softening in a diverse apple germplasm, *Tree Genet. Genomes* 9(1):279–290.

Oraguzie, N. C., Iwanami, H., Soejima, J., Harada, T. and Hall, A. (2004). Inheritance of the *Md-ACS1* gene and its relationship to fruit softening in apple (*Malus* × *domestica* Borkh.), *Theor. Appl. Genet.* 108(8):1526–1533.

Orcheski, B., Parker, R. and Brown, S. (2015). Pale green lethal disorder in apple (*Malus*) is caused by a mutation in the *PHYLLO* gene which is essential for phylloquinone (vitamin K_1) biosynthesis, *Tree Genet. Genomes* 11(6):131.

Osorio, L. F., Gezan, S. A., Verma, S. and Whitaker, V. M. (2021). Independent validation of genomic prediction in strawberry over multiple cycles, *Front. Genet.* 11:1862.

Ould Estaghvirou, S. B., Ogutu, J. O., Schulz-Streeck, T., Knaak, C., Ouzunova, M., Gordillo, A. and Piepho, H. P. (2013). Evaluation of approaches for estimating the accuracy of genomic prediction in plant breeding, *BMC Genomics* 14:860.

Peace, C. P. (2017). DNA-informed breeding of rosaceous crops: Promises, progress and prospects, *Hortic. Res.* 4:17006.

Peace, C. P., Luby, J. J., van de Weg, W. E., Bink, M. C. A. M. and Iezzoni, A. F. (2014). A strategy for developing representative germplasm sets for systematic QTL validation, demonstrated for apple, peach, and sweet cherry, *Tree Genet. Genomes* 10(6):1679–1694.

Picton, S., Gray, J. E. and Grierson, D. (1995). Ethylene genes and fruit ripening. In P. J. Davies (Ed.), *Plant Horm.* Dordrecht, Netherlands: Kluwer Academic Publishers, pp. 372–394.

Plagge, H. H. and Maney, T. J. (1937). Factors influencing the development of soggy break-down in apples, *J. Agric. Res.* 55:739–763.

Porep, J. U., Kammerer, D. R. and Carle, R. (2015). On-line application of near infrared (NIR) spectroscopy in food production, *Trends Food Sci. Technol.* 46(2):211–230.

Prusky, D., Bazak, M. and Ben-Arie, R. (1985). Development, persistence, survival, and strategies for control of thiabendazole-resistant strains of *Penicillium expansum* on pome fruits, *Phytopathology* 75(8):877–882.

Reid, M. and Kalcsits, L. (2020). Water deficit timing affects physiological drought response, fruit size, and bitter pit development for 'Honeycrisp' apple, *Plants (Basel)* 9(7):874.

Rosenberger, D. A. (1997). Recent research and changing options for controlling postharvest decays of apples. In *Proceedings of the Harvesting Handling Storage Workshop, Northeast Regional Agricultural Engineering Service Publication NRAES-112*, Cornell University, Ithaca.

Roth, M., Muranty, H., Di Guardo, M., Guerra, W., Patocchi, A. and Costa, F. (2020). Genomic prediction of fruit texture and training population optimization towards the application of genomic selection in apple, *Hortic. Res.* 7:148.

Ru, S., Hardner, C., Carter, P. A., Evans, K., Main, D. and Peace, C. (2016). Modelling of genetic gain for single traits from marker-assisted seedling selection in clonally propagated crops, *Hortic. Res.* 3:16015.

Ru, S., Main, D., Evans, K. and Peace, C. (2015). Current applications, challenges, and perspectives of marker-assisted seedling selection in Rosaceae tree fruit breeding, *Tree Genet. Genomes* 11(1):8.

Rymenants, M., van de Weg, E., Auwerkerken, A., De Wit, I., Czech, A., Nijland, B., Heuven, H., De Storme, N. and Keulemans, W. (2020). Detection of QTL for apple fruit acidity and sweetness using sensorial evaluation in multiple pedigreed full-sib families, *Tree Genet. Genomes* 16(5):71.

Schlathölter, I., Jänsch, M., Flachowsky, H., Broggini, G. A. L., Hanke, M. V. and Patocchi, A. (2018). Generation of advanced fire blight-resistant apple (*Malus* × *domestica*) selections of the fifth generation within 7 years of applying the early flowering approach, *Planta* 247(6):1475–1488.

Serra, S., Leisso, R., Giordani, L., Kalcsits, L. and Musacchi, S. (2016). Crop load influences fruit quality, nutritional balance, and return bloom in 'Honeycrisp' apple, *Hortscience* 51(3):236–244.

Shewfelt, R. L. (1999). What is quality?, *Postharvest Biol. Technol.* 15(3):197–200.

Song, J. and Forney, C. F. (2008). Flavour volatile production and regulation in fruit, *Can. J. Plant Sci.* 88(3):537–550.

Sunako, T., Sakuraba, W., Senda, M., Akada, S., Ishikawa, R., Niizeki, M. and Harada, T. (1999). An allele of the ripening-specific 1-aminocyclopropane-1-carboxylic acid synthase gene (*ACS1*) in apple fruit with a long storage life, *Plant Physiol.* 119(4):1297–1304.

Takos, A. M., Jaffé, F. W., Jacob, S. R., Bogs, J., Robinson, S. P. and Walker, A. R. (2006). Light-induced expression of a *MYB* gene regulates anthocyanin biosynthesis in red apples, *Plant Physiol.* 142(3):1216–1232.

Taniwaki, M., Hanada, T. and Sakurai, N. (2006). Device for acoustic measurement of food texture using a piezoelectric sensor, *Food Res. Int.* 39(10):1099–1105.

Tazawa, J., Oshino, H., Kon, T., Kasai, S., Kudo, T., Hatsuyama, Y., Fukasawa-Akada, T., Yamamoto, T. and Kunihisa, M. (2019). Genetic characterization of flesh browning trait in apple using the non-browning cultivar 'Aori 27', *Tree Genet. Genomes* 15(4):49.

Teh, S. L., Brutcher, L., Schonberg, B. and Evans, K. (2020a). Eleven-year correlation of physical fruit texture traits between computerized penetrometers and sensory assessment in an apple breeding program, *hortTechnology* 30(6):719–724.

Teh, S. L., Coggins, J. L., Kostick, S. A. and Evans, K. M. (2020b). Location, year, and tree age impact NIR-based postharvest prediction of dry matter concentration for 58 apple accessions, *Postharvest Biol. Technol.* 166:111125.

Teh, S. L., Kostick, S., Brutcher, L., Schonberg, B., Barritt, B. and Evans, K. (2021a). Trends in fruit quality improvement from 15 years of selection in the apple breeding program of Washington State University, *Front. Plant Sci.* 12:714325.

Teh, S. L., Kostick, S. A. and Evans, K. M. (2021b). Genetics and breeding of apple scions. In S. S. Korban (Ed.), *The Apple Genome*. Cham, Switzerland: Springer, pp. 73–103.

Tong, C. B. S., Bedford, D. S., Luby, J. J., Propsom, F. M., Beaudry, R. M., Mattheis, J. P., Watkins, C. B. and Weis, S. A. (2003). Location and temperature effects on soft scald in 'Honeycrisp' apples, *Hortscience* 38(6):1153–1155.

United States Department of Agriculture (USDA). (2011). APHIS confirmation of the regulatory status of null segregant (NS) lines derived from genetically engineered (GE) plants in an accelerated tobacco breeding program. Available at: https://www.aphis.usda.gov/biotechnology/downloads/reg_loi/Dr%20Ramsey%20S%20Lewis%20NCSC%20Final.pdf (accessed 25 April 2022).

Velasco, R., Zharkikh, A., Affourtit, J., Dhingra, A., Cestaro, A., Kalyanaraman, A., Fontana, P., Bhatnagar, S. K., Troggio, M., Pruss, D., Salvi, S., Pindo, M., Baldi, P., Castelletti, S., Cavaiuolo, M., Coppola, G., Costa, F., Cova, V., Dal Ri, A., Goremykin, V., Komjanc, M., Longhi, S., Magnago, P., Malacarne, G., Malnoy, M., Micheletti, D., Moretto, M., Perazzolli, M., Si-Ammour, A., Vezzulli, S., Zini, E., Eldredge, G., Fitzgerald, L. M., Gutin, N., Lanchbury, J., Macalma, T., Mitchell, J. T., Reid, J., Wardell, B., Kodira, C., Chen, Z., Desany, B., Niazi, F., Palmer, M., Koepke, T., Jiwan, D., Schaeffer, S., Krishnan, V., Wu, C., Chu, V. T., King, S. T., Vick, J., Tao, Q., Mraz, A., Stormo, A., Stormo, K., Bogden, R., Ederle, D., Stella, A., Vecchietti, A., Kater, M. M., Masiero, S., Lasserre, P., Lespinasse, Y., Allan, A. C., Bus, V., Chagné, D., Crowhurst, R. N., Gleave, A. P., Lavezzo, E., Fawcett, J. A., Proost, S., Rouzé, P., Sterck, L., Toppo, S., Lazzari, B., Hellens, R. P., Durel, C. E., Gutin, A., Bumgarner, R. E., Gardiner, S. E., Skolnick, M., Egholm, M., Van de Peer, Y., Salamini, F. and Viola, R. (2010). The genome of the domesticated apple (*Malus* × *domestica* Borkh.), *Nat. Genet.* 42(10):833–839.

Verma, S., Evans, K., Guan, Y., Luby, J. J., Rosyara, U. R., Howard, N. P., Bassil, N., Bink, M. C. A. M., van de Weg, W. E. and Peace, C. P. (2019). Two large-effect QTLs, *Ma* and *Ma3*, determine genetic potential for acidity in apple fruit: Breeding insights from a multi-family study, *Tree Genet. Genomes* 15(2):18.

Volz, R. K., Alspach, P. A., Fletcher, D. J. and Ferguson, I. B. (2006). Genetic variation in bitter pit and fruit calcium concentrations within a diverse apple germplasm collection, *Euphytica* 149(1–2):1–10.

Volz, R. K., Alspach, P. A., White, A. G. and Ferguson, I. B. (2001). Genetic variability in apple fruit storage disorders, *Acta Hortic.* 553:241–244.

Wagenmakers, P. S. and Callesen, O. (1995). Light distribution in apple orchard systems in relation to production and fruit quality, *J. Hortic. Sci.* 70(6):935–948.

Wakasa, Y., Kudo, H., Ishikawa, R., Akada, S., Senda, M., Niizeki, M. and Harada, T. (2006). Low expression of an endopolygalacturonase gene in apple fruit with long-term storage potential, *Postharvest Biol. Technol.* 39(2):193–198.

Wang, A., Yamakake, J., Kudo, H., Wakasa, Y., Hatsuyama, Y., Igarashi, M., Kasai, A., Li, T. and Harada, T. (2009). Null mutation of the *MdACS3* gene, coding for a ripening-specific 1-aminocyclopropane-1-carboxylate synthase, leads to long shelf life in apple fruit, *Plant Physiol.* 151(1):391–399.

Wang, H., Peng, J., Xie, C., Bao, Y. and He, Y. (2015). Fruit quality evaluation using spectroscopy technology: A review, *Sensors (Basel)* 15(5):11889–11927.

Wannemuehler, S. D., Luby, J. J., Yue, C., Bedford, D. S., Gallardo, R. K. and McCracken, V. A. (2019). A cost-benefit analysis of DNA informed apple breeding, *Hortscience* 54(11):1998–2004.

Watkins, C. B. (2017). Advances in postharvest handling and storage of apples. In K. Evans (Ed.), *Achieving Sustainable Cultivation of Apples*. Pennsylvania: Burleigh Dodds Science Publishing, pp. 337–368.

Whitaker, V. M., Lee, S., Osorio, L. F., Verma, S., Roach, J. A., Mangandi, J., Noh, Y.-H., Gezan, S. and Peres, N. (2017). Advances in strawberry breeding at the University of Florida, *Acta Hortic.* 1156:1-6.

Wu, B., Shen, F., Wang, X., Zheng, W. Y., Xiao, C., Deng, Y., Wang, T., Yu Huang, Z., Zhou, Q., Wang, Y., Wu, T., Feng Xu, X., Hai Han, Z. and Zhong Zhang, X. (2021). Role of *MdERF3* and *MdERF118* natural variations in apple flesh firmness/crispness retainability and development of QTL-based genomics-assisted prediction, *Plant Biotechnol. J.* 19(5):1022-1037.

Wu, Q., Szakacs-Dobozi, M., Hemmat, M. and Hrazdina, G. (1993). Endopolygalacturonase in apples (*Malus domestica*) and its expression during fruit ripening, *Plant Physiol.* 102(1):219-225.

Xu, K., Wang, A. and Brown, S. (2012). Genetic characterization of the Ma locus with pH and titratable acidity in apple, *Mol. Breed.* 30(2):899-912.

Xu, Y. and Crouch, J. H. (2008). Marker-assisted selection in plant breeding: From publications to practice, *Crop Sci.* 48(2):391-407.

Yang, S. F. and Hoffman, N. E. (1984). Ethylene biosynthesis and its regulation in higher plants, *Annu. Rev. Plant Physiol.* 35(1):155-189.

Yue, C., Gallardo, R. K., Luby, J., Rihn, A., McFerson, J. R., McCracken, V., Bedford, D., Brown, S., Evans, K., Weebadde, C., Sebolt, A. and Iezzoni, A. F. (2013). An investigation of U.S. apple producers' trait prioritization - Evidence from audience surveys, *Hortscience* 48(11):1378-1384.

Zhang, X. J., Wang, L. X., Chen, X. X., Liu, Y. L., Meng, R., Wang, Y. J. and Zhao, Z. Y. (2014). A and *MdMYB1* allele-specific markers controlling apple (*Malus* × *domestica* Borkh.) skin color and suitability for marker-assisted selection, *Genet. Mol. Res.* 13(4):9103-9114.

Zhang, Y., Nock, J. F., Al Shoffe, Y. and Watkins, C. B. (2019). Non-destructive prediction of soluble solids and dry matter contents in eight apple cultivars using near-infrared spectroscopy, *Postharvest Biol. Technol.* 151:111-118.

Zharkikh, A., Troggio, M., Pruss, D., Cestaro, A., Eldrdge, G., Pindo, M., Mitchell, J. T., Vezzulli, S., Bhatnagar, S., Fontana, P., Viola, R., Gutin, A., Salamini, F., Skolnick, M. and Velasco, R. (2008). Sequencing and assembly of highly heterozygous genome of *Vitis vinifera* L. cv Pinot Noir: Problems and solutions, *J. Biotechnol.* 136(1-2):38-43.

Zhen, Q., Fang, T., Peng, Q., Liao, L., Zhao, L., Owiti, A. and Han, Y. (2018). Developing gene-tagged molecular markers for evaluation of genetic association of apple *SWEET* genes with fruit sugar accumulation, *Hortic. Res.* 5:14.

Zheng, W., Shen, F., Wang, W., Wu, B., Wang, X., Xiao, C., Tian, Z., Yang, X., Yang, J., Wang, Y., Wu, T., Xu, X., Han, Z. and Zhang, X. (2020). Quantitative trait loci-based genomics-assisted prediction for the degree of apple fruit cover color, *Plant Genom.* 13(3):e20047.

Zhu, Y. and Barritt, B. H. (2008). Md-ACS1 and Md-ACO1 genotyping of apple (*Malus* × *domestica* Borkh.) breeding parents and suitability for marker-assisted selection, *Tree Genet. Genomes* 4(3):555-562.

Zhu, Y., Evans, K. and Peace, C. (2011). Utility testing of an apple skin color MdMYB1 marker in two progenies, *Mol. Breed.* 27(4):525-532.

Chapter 6

Advances in understanding pre-harvest apple fruit development

Luigi Manfrini and Alessandro Bonora, Bologna University, Italy

1 Introduction

Fruit growth can be defined as the irreversible increase in size of a fruit. Growth is the net balance between incoming and outgoing flows in the fruit. Xylem (transporting water and minerals from the roots) and phloem (transporting sucrose, amino acids and other nutrients from leaves) are the main sources for water and dry matter (DM) transportation/accumulation while epidermal transpiration (both from leaves and fruit) is the main cause of water loss (Fishman and Génard, 1998; Rossi et al., 2022). The varying balance among these fluxes determines typical daily fluctuation in fruit size, controlling fruit development, swelling and shrinkage (Fig. 1.) (Berger and Selles, 1993; Morandi et al., 2007, 2011).

Phloematic and xylematic fluxes influence apple growth almost equally during the first part of the season (i.e. first 60 days after full bloom - DAFB). However, input from xylem flow progressively decreases and, around 90 DAFB, apple growth is sustained by phloem flux (Lang, 1990; Morandi et al., 2011). Hydrostatic pressure gradients along the vascular path are the main drivers of phloem and xylem flows (Patrick, 1997; Rossi et al., 2022), although assimilate

http://dx.doi.org/10.19103/AS.2023.0127.08

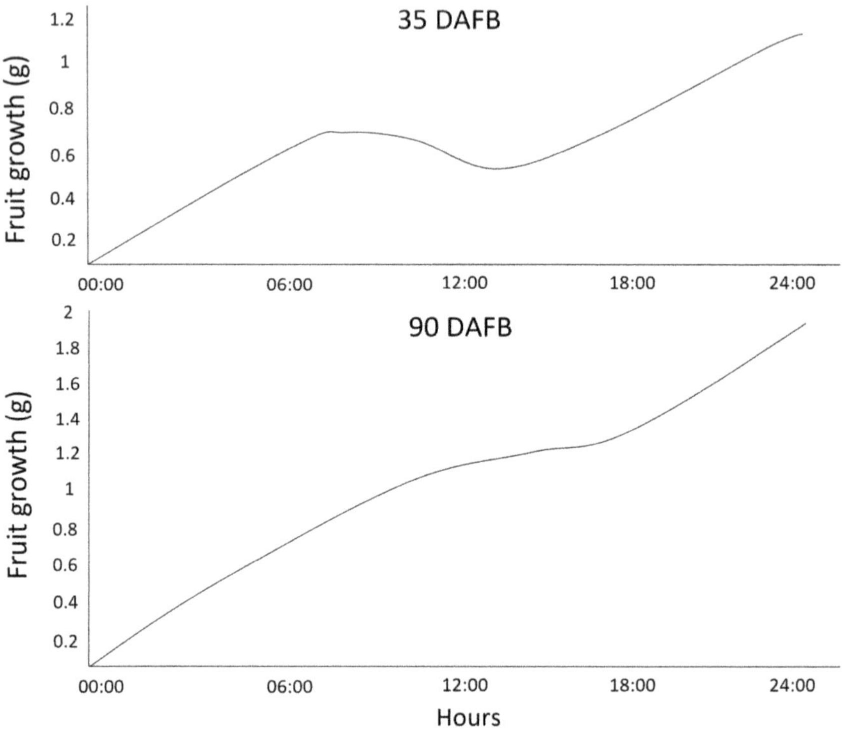

Figure 1 Apple fruit growth: during cytokinesis at 35 DAFB and during cell expansion at 90 DAFB. In the early stages, the fruit tend to shrink during the warmest hours of the day due to high skin transpiration and xylematic backflow.

unloading to apple fruit sink cells occurs through an active process (Zhang et al., 2004; Wei et al., 2014). Apple fruit water losses through the epidermis decrease pressure potential, thus increasing unloading of phloem and xylem sap. Skin transpiration is related to air vapour pressure deficit (VPD) and fruit surface conductance. Conductance is higher in the first stages of fruit growth - although relatively low in apple - but decreases as fruit develop (Jones and Higgs, 1982).

Apples are known to lose lower amounts of water via transpiration in early stages after blooming, compared to other species such as peach (Morandi et al., 2007) or kiwifruit (Torres-Ruiz et al., 2016). This can be related to the environmental conditions typical of their genetic origin and how evolution has provided fruits with high skin permeability to water vapour, providing an efficient hydraulic system to recover transpiration water losses through the xylem (Morandi et al., 2016). During the first stages of development, apples can also lose water via xylem backflow from fruit to leaves (Lang, 1990) as in the second stages xylem vessels become dysfunctional (Dražeta et al., 2004a). Apple fruit growth rates in the first half of the season are likely to be more sensitive to changes in environmental conditions due to higher transpiration rates and xylem flow, as these may easily modify the daily equilibrium between water flows in and out of the fruit.

There are two fundamental processes characterising fruit development:

- Cell division;
- Cell expansion.

These steps significantly influence the growth of the reproductive organs. Their role varies considerably between species or even varieties of the same species. In apple fruit and stone fruit, cytokinesis (cell division) can last up to 35–60 DAFB, while in pears it lasts up to 9 weeks (Westwood, 1978, Karim et al., 2022). Final fruit size is the product of the number of cells and their volume plus the amount of intercellular space (IS). In apple, most cell division events take place in the developing bud before anthesis (Smith, 1950; Harada et al., 2005) with more than 20 cell divisions before flowering, continuing for up to 8 weeks after pollination (Harada et al., 2005; Karim et al., 2022).

The duration of cytokinesis can be directly related to environmental and management factors (Musacchi and Serra, 2018). Smith (1950) found between 3 and 6 million cells in the flower receptacle of nine apple varieties and between 27 and 67 million cells in the cortical parenchyma at harvest. These results are supported by data from Goffinet et al. (1995) and Manfrini (2004) who estimated between 37 and 59 million cells in the fruit pericarp of Empire and Gala cultivars at harvest (Fig. 2). According to these authors, only three to four cell division events are required to achieve the number present in the flower receptacle compared to that of the fruit cortical tissue at harvest. The relatively low number of cell divisions explains the effectiveness of fruit thinning in increasing the size of the fruit. Early thinning (as happens with chemical thinning) has a geometric effect on the number of cells if it corresponds with an increased number of cell divisions.

Several authors have reported the existence of a linear relationships between cell number and fruit weight at harvest (Smith, 1950; Bain and Robertson, 1951; Denne, 1960; Goffinet et al., 1995; Harada et al., 2005; Karim et al., 2022). This led to the conclusion that the number of cells reached at the end of cytokinesis determines maximum potential fruit development. To what extent this potential will be achieved depends on the phase of cell distension. This depends on any growth limitations affecting that phase such as competition from excessive fruit loads or adverse environmental conditions.

This linear relationship, however, tends to flatten out at very high cell counts, as reported by Goffinet et al. (1995) and Harada et al. (2005). For fruits with a high number of cells, weight increase is no longer proportional with cell numbers. This curvilinear pattern suggests a compensatory mechanism

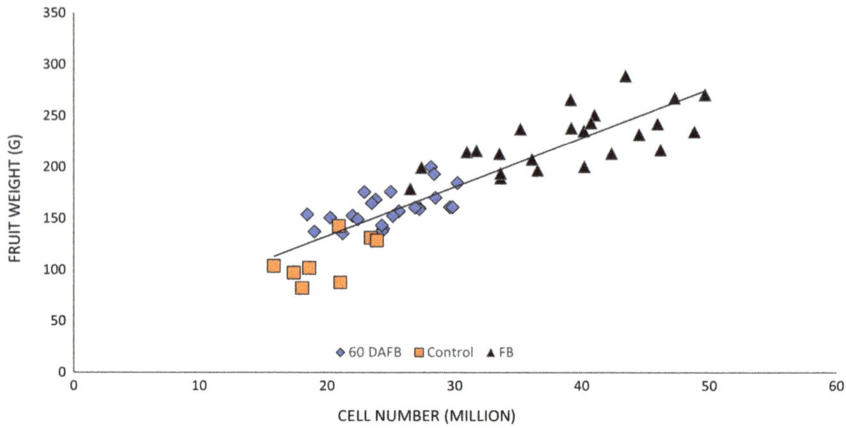

Figure 2 Linear regression of fruit weight plotted against cell number within the cortex for 'Gala' apple fruit at final harvest. Fruit size corresponds to the average size distribution within tree thinning treatments: FB, 60 DAFB and unthinned (control). Source: Modified from: Manfrini (2004).

that reduces cell size as their number increases. The interaction of cytokinesis and cellular distension determines final fruit size, making techniques that can influence these processes particularly important. Research on apple fruit anatomy have, for instance, confirmed that final size is closely associated with the difference in the number of cells present in the cortical area of the fruit (Fig. 2), with cell size having a minor effect, and that these processes are affected by thinning (Goffinet et al., 1995; Manfrini, 2004).

2 Apple fruit growth models and precision orchard management

The physiological aspects of fruit growth discussed in the previous section have been used to develop mechanistic models of fruit crop production (Forsey, 1971). These are usually composed of sub-models which represent the main environmental and physiological mechanistic processes.

The expolinear function developed by Goudriaan and Monteith (1990) is a biological model that describes the development of biomass as a function of light interception. This equation is based on three fundamental parameters:

- The maximum relative growth rate (RGR);
- The maximum absolute growth rate (AGR);
- The time lost, i.e. compared to linear growth from the beginning of fruit development.

There is a strong correlation between the RGR of an apple fruit during cell division (early stages of growth of the fruit) and its final size (Fig. 1). A fruit that demonstrates a higher initial growth rate tends to reach a larger size when harvested (Lakso et al., 1989). This is because, in the early period, variations in the growth rate correspond to variations in cytokinesis. Cytokinesis proceeds with a geometric progression if cells maintain a normal level of growth. This is demonstrated by the slope of the straight line characterising the growth phase which depends on the number of cells formed during the exponential phase and subsequent carbon partitioning (Wünsche and Lakso, 2000; Jing and Malladi, 2020). Techniques stimulating cytokinesis after flowering can increase fruit size at the end of the season. Conversely, too many fruits per plant lead to an excessive demand for resources and nutrients, limiting maximum development at harvest. Lakso et al. (1995) argued that this model is well suited to the description of apple growth if there are no stresses or limiting conditions in terms of nutritional and environmental resources.

Lakso and Johnson (1990) presented the MaluSim apple tree model in 1990. Since then it has been steadily modified and improved (Henton et al., 1999; Doerflinger et al., 2015; Reyes et al., 2020). MaluSim simulates seasonal plant carbon production/utilisation. It is composed of four sub-models estimating:

- whole tree leaf area development and abscission;
- canopy photosynthesis and tree respiration;
- carbon partitioning;
- fruit growth and abscission.

Other advanced models that could be adapted to apple are L-peach (Allen et al., 2005) and CSS.Pear (Sousa et al., 2022). These models are based on widely studied variables as:

- light canopy interception;
- within-day and seasonal thermal data;
- canopy photosynthesis and its efficiency;
- tree metabolism maintenance and growth respiration;
- tree organ dry-weight accumulation resulting from leaf-area development and distribution;
- plant/fruit water losses based on specific crop coefficients (i.e. at each phenological stage), as well as estimating evapotranspiration based on soil water availability.

Current precision orchard management (POM) approaches integrate these models and sub-models to obtain real-time information to evaluate crop performance and to monitor target properties (fruit size, yield, etc.).

Manfrini et al. (2012, 2014) and Bergh and Lötze (2006) have developed a protocol based on periodically measuring fruit diameter to predict orchard productivity and yield. However, time-consuming data acquisition is a bottleneck for practical implementation. To address these measurement issues, Longchamps et al. (2022) summarise a wide range of yield-monitoring technologies including direct and indirect proximal sensors as well as remote measurement techniques. These can be combined with robust models to improve orchard monitoring and management.

3 Apple fruit growth: cell organisation and intercellular spaces

The flesh of an apple is mainly composed of parenchyma tissue criss-crossed by vasculature and intercellular spaces (IS) (Esau, 1977; Mebatsion et al., 2006). The cells that make up the pulp are packed in different ways with different orientations (Smith, 1950; Lapsley, 1992). This phenomenon is known as anisotropy and increases from the outside to the inside of the cortex. In contrast, the characteristic shape of the pome is the result of cell growth that starts from the centre of the fruit outwards. This growth is related to radial expansion and elongation of the parenchyma cells (Bain and Robertson, 1951). Cell size at maturity varies in different areas of the fruit. Cells located immediately below the epidermis (cortical area) are small (50 μm), rounded and randomly oriented. Continuing towards the centre of the fruit, a gradual increase in cell size is seen up to a maximum (200–300 μm).

Moving further towards the core of the fruit, vascular bundles of cells (ca. 300 μm diameter) are organised in an increasing radial pattern, acquiring an ellipsoidal, flattened (spheroidal) and columnar shape. Between these columns are radially elongated spaces up to 3 mm long and 100–200 μm wide. The aspect ratio of the intercellular spaces increases almost linearly from outside to inside, depending on variety (Khan and Vincent, 1990; Drazeta et al., 2004b). The range of cell sizes differs according to apple cultivar with an average of 250 μm, assuming a spherical area of approximately 50 000 μm^2 (McAtee et al., 2009).

During development of the fruit, air spaces tend to increase until harvest (Goffinet and Maloney, 1987; Volz et al., 2002) causing a decrease in the specific weight of the fruit and accounting for much of fruiting volume at harvest (Harker and Ferguson, 1988; Ruess and Stösser, 1993; Mendoza et al., 2010). During development, fruit density decreases significantly while cell packing (i.e. cells per mm^3) remains fairly constant (Westwood et al., 1967; Volz et al., 2002). Intercellular spaces in both the cortex and core tissue have a similar diameter to the diameter of cells, suggesting a lysigenous origin. Spaces probably result from programmed cell death (PCD). PCD can be initiated by low oxygen conditions in the core of the fruit when its diameter and gas diffusion resistance increase during growth (Herremans et al., 2015a). PCD has been observed in fruit under external stress such as suboptimal nutrition and storage conditions (Chen et al., 2014; Zheng et al., 2014). After 9 weeks, the porosity of both the core and the cortex of apples does not change significantly. However, pore size keeps increasing, indicating that lysigenous pore formation occurs during the development phase of cell division. During subsequent cell enlargement, pores and cells essentially grow at the same rate (Herremans et al., 2015a).

The gradual loosening of cells due to middle lamella separation, particularly at positions where three or more cells touch each other, creates schizogenous pores that do not increase porosity but may increase the connectivity of the pore network. Schizogenous pores in apple cortex tissue have been observed (Verboven et al., 2008), though high-resolution x-ray computed tomography (CT) imaging of the cell division developmental phase is required to further explore pre- and postharvest pore formation in apple fruit. As a result of different mechanisms of pore formation, IS will have different sizes, shapes and connectivity.

Visualisation of the skeleton of parenchyma pore spaces has shown that it forms a complicated, highly connected network inside the tissue (Verboven et al., 2008; Mendoza et al., 2007). The parenchyma of an apple cannot therefore be considered homogeneous. Its configuration is influenced not by growth of a single cell but by a collective dynamic which influences differences in size, orientation and organisation of cells and air channels during fruit development. All these aspects, related to the anisotropy and heterogeneity of apple flesh, highlight the variation and complexity of parenchymatic tissue development of fruit.

A number of methods have been developed to study pore number and arrangement as well as air content in apples and pears (Fig. 3). Calbo and Sommer (1987) estimated that air spaces comprised around 17% of the total volume of tissue by evacuating the internal gases from the flesh with a vacuum. By replacing IS with water, Dražeta et al. (2004b) showed that the concentration of air spaces tends to decrease in a centripetal pattern, starting from the area immediately below the epidermis. There is an increase of cellular organisation with a radial orientation. This increases from the cortex to the core of the pulp. This is accompanied by increasing cellular organisation following the decrease in airspace. The airspace initially represents about 18% of the cortical area reaching up to 12% of the core. In contrast, Goffinet et al. (1995) suggested an IS range between 32% and 35%. Chigwaya et al. (2021) have measured the total porosity of 'Fuji' apples using x-ray CT and 3D image analysis in samples, both with internal browning after storage (4% of IS) and in unaffected samples (11.9% of IS).

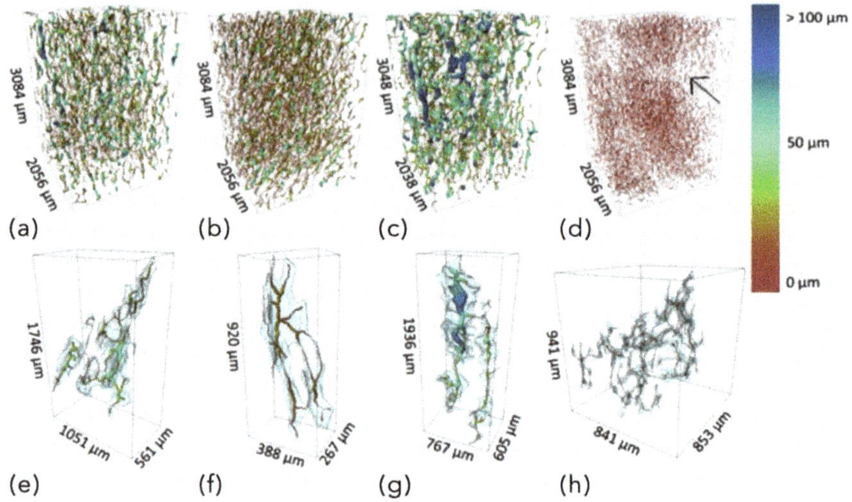

Figure 3 Void network models for (a) 'Braeburn', (b) 'Kanzi' (b), (c) 'Jonagold' and (d) 'Conference' showing void topology and void network branching as well as local thicknesses of the voids expressed by the colour scale. The arrow in the 'Conference' image indicates the presence of stone cells, with a local absence of surrounding voids. Details of the void network models for a single void, show the original void volume (transparent blue) and the calculated void network. For these models, large void volumes were chosen to illustrate the void connectivity. The dimensions of the box illustrate the spatial extent of the void network. Plots (e)–(h) present a single connected void in each of the corresponding void networks in plots (a)–(d), demonstrating variations in size and connectivity. Source: Extracted from Herremans et al. (2015b).

4 Environmental factors affecting fruit growth: light, temperature, altitude and latitude

As for the majority of cultivated fruit crops, key variables affecting apple fruit growth are genetic, biological and crop management factors. However, ecophysiological conditions are also important. Understanding fruit development across the growing season, as affected by environmental factors, is key to effective orchard management.

4.1 Light

Key quality parameters such as size, soluble solids content and colour development of the epidermis are improved from exposure to high levels of light (Jackson et al., 1977). The productivity of apple orchards is known to be limited by low light interception (Tustin et al., 2018). Several studies have shown that the upper limit of light interception by mature apple orchards is approximately 60% of ambient incoming radiation (Wünsche et al., 1996; Palmer et al., 2002). Light reduction produced by intense shading during early fruit growth stages decreases apple fruit growth rate and causes fruit drop (McArtney et al., 2004). This is directly associated with reductions in photosynthesis and tree carbon assimilation (Morandi et al., 2011).

Light also affects carbohydrate partitioning between fruit and shoots. Early in the season, when sunlight is limited, shoot growth is a stronger sink for photo-assimilate allocation (Bepete and Lakso, 1998; Corelli-Grappadelli, 2003). However, mild light intensity limitation (i.e. up to 40–50%) does not alter apple tree physiological development and productivity even in water limited conditions (Boini et al., 2022, 2023; Lopez et al., 2018). Trees grown under shade nets had a better water status than trees grown in full light, indicating lower irrigation requirements. Photosynthetic activity of leaves is not reduced by shading, probably due to the beneficial effect of reducing excessive photon pressure but also possibly due to morphological and physiological adaptation – i.e. bigger thinner leaves and more chlorophyll B and optimised xanthophyll ratios for light capture (Close et al., 2001; Tozzi et al., 2018). The combination of improved water status combined with maintenance of leaf photosynthetic performance resulted in higher yields for trees grown under nets.

However, lower light levels can delay fruit ripening, carbohydrate accumulation and harvest dates (Lopez et al., 2018; Losciale et al., 2011). Fruit growers should take this delay into account, considering higher apple prices earlier in the season. Some studies have assessed the use of different colour photo-selective nets (Boini et al., 2019, 2021; Bastías et al., 2012; Kalcsits et al., 2018). Bastías et al. (2012) found greater fruit size under blue compared to red nets, probably due to the positive effect of blue light on leaf photosynthesis, improving net tree carbon assimilation rate and the total DM production. Changing red, blue and far-red light composition using nets could be a useful way to control photosynthetic and morphogenetic processes regulating carbohydrate availability for apple fruit growth.

4.2 Temperature

Climate change is expected to constrain production in traditional apple cultivation regions. Temperature is becoming a more important factor affecting fruit development. Controlled environment studies conducted by Warrington et al. (1999) demonstrated that diurnal fluctuation between maximum and minimum temperature in the early part of the season (first 40 days after full bloom) had a strong effect on apple development. Fruit exposed to average temperatures of 20°C resulted in ten times growth than those exposed to temperatures averages of 6°C. However, low temperatures probably prolonged cell division over time. Stanley et al. (2000) and Flaishman et al. (2015) studied the correlation between degree-days accumulation and fruit development during the season. Their findings indicated that, by regulating cytokinesis, temperature-driven growth in early fruit development appears to be the major factor establishing fruit size and consequent yield potential. They found that lower growth in apple fruitlets at 29°C caused a significant temporary decline in cell production. The expression of several expansion genes has been associated with high temperatures, making them potentially useful as molecular markers to promote cell-expansion processes under high-temperature regimes.

Different fruit growth models have described the effects of tree architecture on fruit thermal responses (Saudreau et al., 2011; Reyes et al., 2016). These models establish a link between fruit position in the canopy and temperature variability in the orchard. New sensors have been developed to provide real-time information on fruit surface temperature from 3D thermal tree point clouds (Tsoulias et al., 2022). The length of key phenological stages also changes according to location. This was initially found in a three-site experiment (Corelli Grappadelli and Lakso, 2005) between Italy (Bologna area), New Zealand, (Hawke's Bay area) and Washington State (Wenatchee area) by analysing fruit growth using the expolinear model (Lakso et al., 1995). Understanding the morpho-physiological mechanisms of heat tolerance associated with cytokinesis and cell expansion may help to develop innovative strategies in breeding of new apple cultivars and improving fruit productivity under climate change.

4.3 Altitude and latitude

Altitude and latitude are two of the most important pre-harvest factors affecting fruit development, as they change the quality and intensity of light, temperature, humidity and precipitation (Eccher and Noe, 1993; Luo et al., 2014; Sahu et al., 2020). These variables affect physiological mechanisms of fruit growth (Cepeda et al., 2021; Argenta et al., 2022). Charles et al. (2018) found that fruit from lower altitude had a higher number of cells, intercellular spaces and a larger total volume. Low altitude conditions promote more cytokinesis and a higher number of cells compared to fruit from higher altitude. This could also enhance fruit growth at lower altitudes. Lower altitudes produced bigger fruit with a higher number of cells per fruit (Harada et al., 2005).

5 Agronomic factors affecting fruit growth: crop load and thinning, irrigation and tree architecture

The main goal of a grower is a high-yield and high-quality crop to satisfy the consumer. There are a large number of agronomic factors that can influence fruit production, many of which are within the control of the grower or are influenced by climatic and soil conditions (which can then be taken into account in planting and management).

5.1 Crop load and thinning

Bain and Robertson (1951) were the first researchers connecting the cellular composition of an apple with different fruit growth patterns. Number and size of cells, and percentage of intercellular spaces, can be stimulated by common practices such reducing crop load. Thinning has now become one of the most important orchard practices. However, the timing of thinning is still less well understood. As previously discussed, cell number is the main factor in obtaining large fruit (Goffinet et al., 1995). Early cell division is influenced by crop load and thinning because carbon is less limited (Lakso et al., 1995; Warrington et al., 1999).

Embree et al. (2007) have highlighted practices to control crop load such as winter pruning, removal of flower buds, flower and fruit thinning during the first stages of fruit development. The main aim of these operations is to leave an appropriate number of fruits per plant for balanced carbon input and optimal fruit size, yield and quality. Manfrini et al. (2015, 2019) undertook a 4-year case study on improving precision in thinning management. They showed it was possible to provide regular feedback to growers about potentially excessive crop load and when best to thin. The removal of fruit in mid-July (late for the growing area) effectively increased fruit growth to reach a satisfactory fruit size at harvest. New machine learning approaches are now showing potential for estimating tree crop load and fruit size with high precision and accuracy (Bresilla et al., 2019; Gené-Mola et al., 2023).

5.2 Irrigation

Drought and heat stress can have negative impacts on the whole supply chain (Stöckle et al., 2011; Boini et al., 2019). These stresses reduce fruit growth and consequently fruit size and yield at harvest. Precision irrigation techniques and models have being developed to control plant water stress at critical growth stages by applying the amount of water needed (Casadesús et al., 2012; Lopez et al., 2018). Precision irrigation requires integration of sensors in a decision support system to efficiently control timing, rate and distribution of water (Jiang and He, 2021). Naor et al. (2008) compared crop yields and fruit sizes of a 'Golden Delicious' orchard with different irrigation schedules, identifying midday stem water threshold potentials as a management parameter for irrigating apple orchards. Lakso et al. (2022) used a stem-embedded microchip micro-tensiometer sensor for continuous, long-term monitoring of stem water potential. Fallahi et al. (2010) undertook a long-term experiment of deficit irrigation of 'Fuji' apple to optimise yields, showing a short-term increase in yield but with lower yields in following years, with larger fruit size when deficit irrigation was applied to only one side of the tree.

Remote sensing spectral reflection techniques have been found effective in assessing plant water potential compared to direct measurement (Gautam and Pagay, 2020). However, since orchards demonstrate large temporal and spatial variability, precision irrigation needs to combine direct measurement using plant and environmental sensors (Scalisi et al., 2017), combined with detailed orchard mapping, remote sensing and modelling (Manfrini et al., 2020). This is an innovative field of research aiming to analyse complex amounts of data using machine learning approaches (Gautam and Pagay, 2020; Tosin et al., 2022).

5.3 Tree architecture

Orchard productivity and total light interception are closely related for many perennial species such as apple (Jackson and Palmer, 1972; Wünsche and Lakso, 2000; Tustin et al., 2018, 2022). An orchard involves considerations such as tree spacing and layout, rootstock-scion interaction, as well as pruning methods (Ferree and Warrington, 2003). Training systems have evolved from a 3D single leader or mono-axis tree to a tall thin fruiting wall, variously named a planar cordon, Guyot tree, multi-axis or multileader system (Dorigoni, 2016;

Figure 4 Examples of multi-axis training system on fourth leaf Pink Lady®-Rosy Glow apple trees, grafted on Pajam 2 rootstock and grown in Po valley (Ferrara) - Italy. (Courtesy of Luigi Manfrini).

Tustin and Van Hooijdonk, 2016; Tustin et al., 2022; Dallabetta et al., 2021; Musacchi et al., 2021; Lezzer et al., 2022; Bortolotti et al., 2022) (Fig. 4).

The general aim is to increase light interception and decrease shelf-shading to directly improve yield and fruit quality. Using a planar cordon spaced 3 m between plants and 1.5 m/2.0 m between-row spacing, a long-term New Zealand experiment found an annual yield increase related to annual maximum fractional light interception from two year onwards, suggesting potential productivity of 169 t ha^{-1} at 90% light interception in spindle systems (Palmer et al., 2002). The theoretical maximum annual yield of market-quality apples at ~90% light utilisation has estimated to be in a range of 250–315 t ha^{-1} (Tustin et al., 2022). These architectures provide thinner and simpler canopies architecture which also facilitate automation and mechanisation (Zhang, 2017; Bortolotti et al., 2021).

6 Fruit development and pre-harvest fruit sensory quality

Pre-harvest effects on the sensory quality of apples and pears have been widely discussed (Bonora et al., 2023; Hudina and Štampar, 2002; Policarpo et al., 2006). These have shown clearly how quality can be affected by environmental factors and orchard management (Rapparini et al., 2008). Using the multivariate partial least squares regressions (PLS) technique, Casero et al. (2004) e.g. found correlations between fruit quality attributes, such as fruit acidity and firmness, with pre-harvest uptake of nutrients such as Ca, K and P, both in leaves and fruit.

Fruit firmness is one of the most important fruit quality indices (Saquet, 2019). Softer fruit have rounder cells separated by larger intercellular spaces than firmer fruit. Firm fruits have denser tissues (smaller cells with less interspace) and can be stored for longer than soft fruits (Johnston et al., 2002). Fruit firmness depends on many factors such as crop load, cultivar, location and storage technique. In apple trees with high crop loads, individual fruits compete for a limited carbohydrate supply. Insufficient carbohydrate reserves limit cell wall development, increasing volume of intercellular space which consequently affects DM and firmness. Trees with low crop loads produce fruit with higher DM and firmness (Link, 2000).

Fruit ripening can be predicted by starch degradation which can be modelled using multivariate statistical techniques (Zude-Sasse et al., 2002). Lau (1985) found that differences in starch patterns start to appear 2–3 weeks before the climacteric peak of ethylene. An optimal starch pattern index (SPI) at harvest suggests the possibility of storing fruit longer (Kingston, 1992). Starch degradation in apple and pear can be influenced by several pre-harvest factors. Environmental factors such as temperature may influence the kinetics of starch accumulation and degradation (Watkins et al., 1982). Deficit irrigation in 'Conference' pear reduced tree organ dry biomass and starch content (Lopez et al., 2013). Conversion of starch to sugars in fruit can be enhanced by water stress (Kramer, 1983). The harvest date also can also affect starch patterns. Starch degradation is slower during on-tree ripening given the continuous source-to-sink supply of carbohydrates occurring in fruit attached to the tree for respiration and other catabolic processes (Lindo-García et al., 2019).

Acidity is a fundamental quality trait (Mattheis and Fellman, 1999) derived from organic acids such as malic acid, citric acid and tartaric acid (Kader, 2008). Malic acid accounts for the 90% of organic acids in apples (Bai et al., 2015). Acidity (at 0.7% of malic acid) is recommended as a parameter for a long-term storage (Rutkowski et al., 2008). Organic acids facilitate represent a substrate for fruit respiration both on and off the tree (Hulme et al., 1963) with malic acid particularly important for harvested apples (Ackermann et al., 1992). Titratable acidity in fruit is influenced by pre-harvest factors such as cultivar (Policarpo et al., 2006), mineral nutrition (Casero et al., 2004; Kader, 2008), low iron (Fe) availability (Álvarez-Fernández et al., 2003), crop load (De Salvador et al., 2006), water availability (Gelly et al., 2004), light exposure (Tombesi et al., 1993; Zhang et al., 2016) and canopy position (Cronje et al., 2015).

Moreover, ammonium ions are the main regulators of cellular pH and the principal precursors for amino acids. Cells react to a high concentration of ammonium ions by accumulating or neutralising organic acids, mediating synthesis of amino acids (Gerendás and Ratcliffe, 2000). High nitrogen availability frequently leads to greater protein levels in food (Rembiałkowska, 2007). Under conditions of light and nutrient deficiency, plants prefer to neutralise excess ammonium via acid accumulation as an alternative to synthesis of amino acids (Bertazza et al., 2010). During storage this biological pathway can lead to increased fumaric acid concentration under reduced respiratory activity. Fumaric acid also has been shown to function as an activator of malate dehydrogenase (Grissom et al., 1983). Fumaric acid is a by-product of the urea cycle which serves to eliminate excess nitrogen due to proteolysis and subsequent amino acid degradation (Pedreschi et al., 2009).

Total sugar content is a key fruit quality trait (Osorio and Fernie, 2014). Total soluble solids (TSS) content is influenced by pre-harvest factors such as nutrition and mineral content, fruit position and light microclimate, picking day and storage techniques (Jajo et al., 2014). Quality traits such as the soluble solids are positively correlated to fruit potassium (K) content (Brunetto et al., 2015). Hamadziripi et al. (2014) found that outer canopy apple fruit have higher TSS and higher antioxidant content compared to inner canopy apples. Kappel and Neilsen (1994) and Zhang et al. (2016) found that 'd'Anjou' soluble solids content at harvest was positively correlated to the percentage of light penetration. Orchards with more exposure to sunlight produce better colour (Brooks and Cooley, 1923) and increased anthocyanins in fruit (Jackson and

Sharples, 1971) with improved resistance to storage disorders (Ramokonyane et al., 2016).

Fruit DM has recently emerged as a fruit quality index (Travers et al., 2014a). DM has been found to improve consumer acceptability of fruit (Palmer et al., 2010). DM is affected by photosynthates and other constituents in fruit tissue (including sugars, starches, proteins, minerals, fibres and acids) acquired during development and which remain once water is removed (Suni et al., 2000). Some authors measure apple DM at or before harvest to predict future internal quality during storage when starch is converted into sugars (McGlone et al., 2003).

The percentage of DM depends on factors such as cultivar, row spacing, pollination, nutrition, water availability, crop load and harvest date. Crop load can affect carbon allocation and consequently DM content (Palmer et al., 2010). Trees with low crop loads produce fruit with higher DM (Link, 2000). DM has been found to be significantly higher in fruit from lower crop loads, with a difference in DM content of 2.8% between fruit from the lowest and the highest crop loads (Fig. 5). After 6 months of storage, DM% showed a 1.9% difference between the two crop loads, being significantly higher in fruit from lower crop loads, compared to fruit from trees with crop loads greater than 11.3 fruit/cm^2 (Serra et al., 2016).

DM is distributed differently e.g. with 22.4% in the skin and 15% in the flesh tissue, with an identical concentration in the outer and inner flesh tissues in the fruit of 'Rocha' pear (Saquet et al., 2019). Fruit physiology may also affect

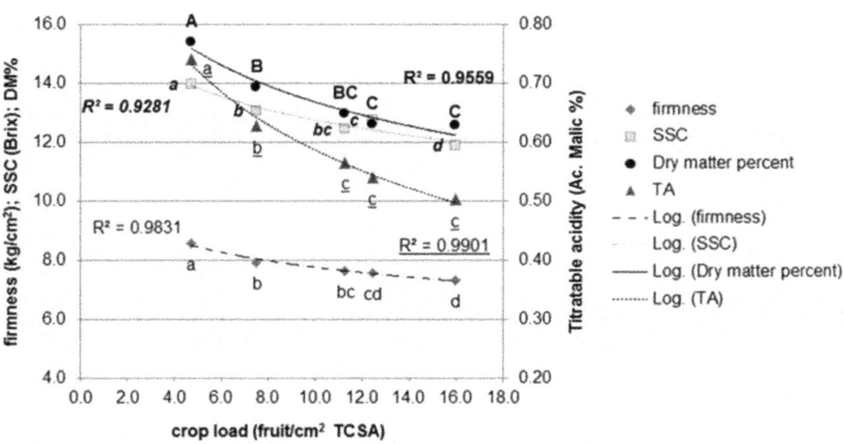

Figure 5 Effect of crop load on fruit quality parameters and maturity such as firmness, soluble solid concentration (SSC), DM%, titratable acidity (TA) shortly after harvest (1 month of storage at 1°C) in 'Honeycrisp' apples grown in the northwest United States. Means comparisons were established with post hoc Tukey test. Means followed by the same letter and font are not statistically different at $P < 0.05$. Source: Extracted by Serra et al. (2016).

DM and SSC given the presence of stone cells just below the skin. These cells contain lignin and are a source of soluble solids in ripening fruit (Travers et al., 2014b).

Water stress can result in fruit with higher DM percentages and better quality traits (Miller et al., 1998). The greater reduction in fruit xylem inflow compared to fruit phloem inflow is the key determinant of higher DM% and soluble solids content typical of drought stressed fruit (Morandi et al., 2014). DM partitioning was also found to be influenced by in-row spacing and cultivar e.g. 'William' pear trees at closer spacings partitioned less DM to roots than 'Conference' pear trees at wider spacings in water stress conditions (Policarpo et al., 2006).

Unfertilised fruit usually shows lower DM than insect-pollinated fruit (Fountain et al., 2019). Many studies have identified seed set as an important indicator of fruit development that influences fruit DM and shape (Garratt et al., 2014). Misshapen apple and pear fruit may be associated with low seed number or flat seeds in the fruit. Poor pollination may also reduce calcium status and increase the occurrence of storage disorders (Brookfield et al., 1996).

7 Conclusion

The growth of fruit is a complex set of processes. The eco-physiological understanding of these processes is fundamental to increasing production and maintaining the highest fruit quality standards. Apple growth potential is determined by cytokinesis: a higher cell number produces larger fruit. However, final fruit size is affected by successive phases of growth, including cell distension and consequent formation of intercellular spaces. The extent of cell expansion also depends on absence of limitations in availability of nutrients, water and carbon to support fruit growth. These factors can cause imbalances in supply of carbon to fruit, resulting in reduced growth. The most important environmental factors affecting fruit growth are light and temperature. At the beginning of the fruiting season, temperature is the main factor in determining the size of the fruit because of its effect in stimulating cell division. Techniques such as covering orchards with plastic films, photo-selective nets and agri-voltaic systems allow the thermal conditioning of orchards and protect from late frost events. Light, its interception and distribution in the canopy, are influenced by training systems and pruning practices and are key given the direct relationship between light interception and yield. Agronomic and environmental factors such as crop load, light diffusivity, temperature and irrigation in different phenological stage during the growing season affect fruit maturity and quality traits, influencing apple storage and consumer acceptance. Carbon availability and water supply are able to modify apple fruit firmness, acidity and sugar content. Integrating understanding of the mechanisms behind fruit growth and quality with adoption of precision orchard management techniques, using real-time innovative solutions, will help to maintain and increase the profitability of pome production.

8 Where to look for further information

A comprehensive introduction to the topic of pre-harvest factors influencing apple fruit development can be found in *Apples: botany, production and uses* by Ferree and Warrington (2003). For the most extensive information on apple fruit development, the International Society for Horticultural Science (ISHS) website at www.ishs.org is the definitive source. ISHS serves as the foremost independent organization for horticulturists globally, serving as a widely recognized and highly sought-after platform for the exchange of science-based research and collaboration in support of sustainable innovation in horticulture. The society also publishes the Acta Horticulturae collections, which are accessible online at www.actahort.org.

9 References

Ackermann, J., Fischer, M. and Amadò, R. (1992). Changes in sugars, acids, and amino acids during ripening and storage of apples (cv. Glockenapfel). *J. Agric. Food Chem.* 40(7), 1131–1134.

Allen, M. T., Prusinkiewicz, P. and DeJong, T. M. (2005). Using L-systems for modeling source-sink interactions, architecture and physiology of growing trees: the L-PEACH model. *New Phytol.* 166(3), 869–880.

Álvarez-Fernández, A., Paniagua, P., Abadía, J. and Abadía, A. (2003). Effects of Fe deficiency chlorosis on yield and fruit quality in peach (Prunus persica L. Batsch). *J. Agric. Food Chem.* 51(19), 5738–5744.

Argenta, L. C., do Amarante, C. V. T., de Freitas, S. T., Brancher, T. L., Nesi, C. N. and Mattheis, J. P. (2022). Fruit quality of 'Gala' and 'Fuji' apples cultivated under different environmental conditions. *Sci. Hortic. (Amsterdam)* 303, 111195.

Bai, Y., Dougherty, L., Cheng, L., Zhong, G. Y. and Xu, K. (2015). Uncovering co-expression gene network modules regulating fruit acidity in diverse apples. *BMC Genomics* 16(1), 612.

Bain, J. M. and Robertson, R. N. (1951). The physiology of growth in apple fruits. I. Cell size, cell number, and fruit development. *Aust. J. Sci. Res. B* 4(2), 75–107.

Bastías, R. M., Manfrini, L. and Grappadelli, L. C. (2012). Exploring the potential use of photo-selective nets for. *Chil. J. Agric. Res.* 72, 224–231.

Bepete, M. and Lakso, A. N. (1998). Differential effects of shade on early-season fruit and shoot growth rates in "Empire" apple. *Hortscience* 33(5), 823–825.

Berger, A. and Selles, G. (1993). Diurnal fruit shrinkage: a model, 261–267. In: Borghetti, M., Grace, J. and Raschi, A. (Eds). *Water Transport in Plants under Climatic Stress*. Cambridge, UK: Cambridge University Press.

Bergh, O. and Lötze, E. (2006). Mid season prediction of fruit size distribution of "Golden Delicious" apples in South Africa. *Acta Hortic.* 707, 217–222.

Bertazza, G., Cristoferi, G. and Bignami, C. (2010). Fruit composition and quality of organically and conventionally grown apple, apricot and pear in the Veneto region (Northern Italy). *Acta Hortic.* (873), 309–316.

Boini, A., Bortolotti, G., Perulli, G. D., Venturi, M., Bonora, A., Manfrini, L. and Corelli Grappadelli, L. (2023). Gala apple production benefits from high shading levels and water limitation, under exclusion netting. *Sci. Hortic.* 310, 111756.

Boini, A., Bortolotti, G., Perulli, G. D., Venturi, M., Bonora, A., Manfrini, L. and Corelli-Grappadelli, L. (2022). Late ripening apple production benefits from high shading and water limitation under exclusion netting. *Horticulturae* 8(10), 111756.

Boini, A., Bresilla, K., Perulli, G. D., Manfrini, L., Corelli Grappadelli, L. and Morandi, B. (2019). Photoselective nets impact apple sap flow and fruit growth. *Agric. Water Manag.* 226, 105738.

Boini, A., Muzzi, E., Tixier, A., Zwieniecki, M., Manfrini, L. and Grappadelli, L. C. (2021). Photoselective nets alter apple canopy air temperature and carbon translocation during dormancy and budbreak. *Hort.Sci.* 56(10), 1166–1174.

Bonora, A., Venturoli, A., Venturi, M., Boini, A. and Corelli Grappadelli, L. (2023). Fruit maturity and antioxidant activity affecting superficial scald development in 'Abate Fétel' pears. *Adv. Hortic. Sci.* 37(1), 3–13.

Bortolotti, G., Bresilla, K., Piani, M., Grappadelli, L. C. and Manfrini, L. (2021). 2D tree crops training system improve computer vision application in field: A case study. In: 2021 IEEE International Workshop on Metrology for Agriculture and Forestry, MetroAgriFor 2021 – Proceedings, 120–124.

Bortolotti, G., Perulli, G., Boini, A., Bresilla, K., Bonora, A., Venturi, M. and Manfrini, L. (2022). Individual upright physiological traits in an apple "Guyot" training system. *Acta Hortic.* 1346, 353–358.

Bresilla, K., Perulli, G. D., Boini, A., Morandi, B., Corelli Grappadelli, L. and Manfrini, L. (2019). Single-shot convolution neural networks for real-time fruit detection within the tree. *Front. Plant Sci.* 10, 1–12.

Brookfield, P. L., Ferguson, I. B., Watkins, C. B. and Bowen, J. H. (1996). Seed number and calcium concentrations of "Braeburn" apple fruit. *J. Hortic. Sci. Biotechnol.* 71(2), 265–271.

Brooks, C., Cooley, J. S. and Fischer, D. F. (1923). Apple scald and its control. *Farmers' Bull* 1380, 83–205.

Brunetto, G., Melo, G. W. B. D. De, Toselli, M., Quartieri, M. and Tagliavini, M. (2015). The role of mineral nutrition on yields and fruit quality in grapevine, pear and apple. *Rev. Bras. Frutic.* 37(4), 1089–1104.

Calbo, A. G. and Sommer, N. F. (1987). Intercellular volume and resistance to mass air flow of fruits and vegetables. *J. Amer. Soc. Hort. Sci.* 112(1), 131–134.

Casadesús, J., Mata, M., Marsal, J. and Girona, J. (2012). A general algorithm for automated scheduling of drip irrigation in tree crops. *Comput. Electron. Agric.* 83, 11–20.

Casero, T., Benavides, A., Puy, J. and Recasens, I. (2004). Relationships between leaf and fruit nutrients and fruit quality attributes in Golden Smoothee apples using multivariate regression techniques. *J. Plant Nutr.* 27(2), 313–324.

Cepeda, M., Vélez-Sánchez, J. E. and Balaguera-López, H. E. (2021). Analysis of growth and physicochemical changes in apple cv. Anna in a high-altitude tropical climate. *Rev. Colomb. Cienc. Hortic.* 15(2), 0–3.

Charles, M., Corollaro, M. L., Manfrini, L., Endrizzi, I., Aprea, E., Zanella, A., Corelli Grappadelli, L. and Gasperi, F. (2018). Application of a sensory-instrumental tool to study apple texture characteristics shaped by altitude and time of harvest. *J. Sci. Food Agric.* 98(3), 1095–1104.

Chen, J., Zhao, Y., Chen, X., Peng, Y., Hurr, B. M. and Mao, L. (2014). The role of ethylene and calcium in programmed cell death of cold-stored cucumber fruit. *J. Food Biochem.* 38(3), 337–344.

Chigwaya, K., Plessis, A. du, Viljoen, D. W., Crouch, I. J. and Crouch, E. M. (2021). Use of X-ray computed tomography and 3D image analysis to characterize internal browning in 'Fuji' apples after exposure to CO2 stress. *Sci. Hortic.* 277, 109840.

Close, D. C., Davies, N. W. and Beadle, C. L. (2001). Temporal variation of tannins (galloylglucoses), flavonols and anthocyanins in leaves of Eucalyptus nitens seedlings: implications for light attenuation and antioxidant activities. *Functional Plant Biol.* 28(4), 269–278.

Corelli-Grappadelli, L. (2003). Light relations. In: Ferree, D. C. and Warrington, I. J. (Eds). *Apples: Botany, Production and Uses*. Wallington, Oxford, UK: CAB International, 195–216.

Corelli Grappadelli, L. and Lakso, A. N. (2005). L'effetto di fattori fisiologici e delle condizioni ambientali sullo sviluppo dei frutti. *Italus Hortus* 12(19), 32.

Cronje, A., Crouch, E. M., Muller, M., Theron, K. I., van der Rijst, M. and Steyn, W. J. (2015). Canopy position and cold storage duration affects mealiness incidence and consumer preference for the appearance and eating quality of "Forelle" pears. *Sci. Hortic.* 194, 327–336.

Dallabetta, N., Guerra, A., Pasqualini, J. and Fazio, G. (2021). Performance of semi-dwarf apple rootstocks in two-dimensional training systems. *Hortscience* 56(2), 234–241.

De Salvador, F. R., Fisichella, M. and Fontanari, M. (2006). Correlations between fruit size and fruit quality in apple trees with high and standard crop load levels. *J. Fruit Ornam Plant Res.* 14, 113-122.

Denne, M. P. (1960). The growth of apple fruitlets, and the effect of early thinning on fruit development. *Ann. Bot. Land*. 24, 597–406.

Doerflinger, F. C., Lakso, A. N. and Braun, P. (2015). Adapting the MaluSim apple tree model for the "Gala" cultivar. *Acta Hortic.* 1068, 267–272.

Dorigoni, A. (2016). Innovative fruit tree architecture as a nexus to improve sustainability in orchards. *Acta Hortic.* 1137, 1–10.

Dražeta, L., Lang, A., Hall, A. J., Volz, R. K. and Jameson, P. E. (2004a). Causes and effects of changes in xylem functionality in apple fruit. *Ann. Bot.* 93(3), 275–282.

Dražeta, L., Lang, A., Hall, A. J., Volz, R. K. and Jameson, P. E. (2004b). Air volume measurement of "Braeburn" apple fruit. *J. Exp. Bot.* 55(399), 1061–1069.

Eccher, T. and Noe, N. (1993). Influence of light on shape and quality of Golden Delicious apples. *Acta Hortic.* 329, 156–158.

Embree, C. G., Myra, M. T. D., Nichols, D. S. and Wright, A. H. (2007). Effect of blossom density and crop load on growth, fruit quality, and return bloom in "Honeycrisp" apple. *Hortscience* 42(7), 1622–1625.

Esau, K. (1977). *Anatomy of Seed Plants* (2nd edn.). New York: John Wiley & Sons.

Fallahi, E., Neilsen, D., Neilsen, G. H., Fallahi, B. and Shafii, B. (2010). Efficient irrigation for optimum fruit quality and yield in apples. *Hort.Sci.* 45(11), 1616–1625.

Ferree, D. C. and Warrington, I. J. (2003). *Apples: Botany, Production and Uses*. Wallingford: CABI Publishing.

Fishman, S. and Génard, M. (1998). A biophysical model of fruit growth: simulation of seasonal and diurnal dynamics of mass. *Plant Cell & Environment* 21(8), 739–752.

Flaishman, M. A., Peles, Y., Dahan, Y., Milo-Cochavi, S., Frieman, A. and Naor, A. (2015). Differential response of cell-cycle and cell-expansion regulators to heat stress in apple (Malus domestica) fruitlets. *Plant Sci.* 233, 82–94.

Forsey, C. G. (1971). Predicting harvest size of McIntosh apples. *N. Y.S Food Life Sci. Bull.* 9, 1-8.

Fountain, M. T., Mateos-Fierro, Z., Shaw, B., Brain, P. and Delgado, A. (2019). Insect pollinators of conference pear (Pyrus communis L.) and their contribution to fruit quality. *J Pollinat Ecol.* 25, 103–114.

Garratt, M. P. D., Breeze, T. D., Jenner, N., Polce, C., Biesmeijer, J. C. and Potts, S. G. (2014). Avoiding a bad apple: insect pollination enhances fruit quality and economic value. *Agric. Ecosyst. Environ.* 184(100), 34–40.

Gautam, D. and Pagay, V. (2020). A review of current and potential applications of remote sensing to study the water status of horticultural crops. *Agronomy* 10(1), 140.

Gelly, M., Recasens, I., Girona, J., Mata, M., Arbones, A., Rufat, J. and Marsal, J. (2004). Effects of stage II and postharvest deficit irrigation on peach quality during maturation and after cold storage. *J. Sci. Food Agric.* 84(6), 561–568.

Gené-Mola, J., Ferrer-Ferrer, M., Gregorio, E., Blok, P. M., Hemming, J., Morros, J. R., Rosell-Polo, J. R., Vilaplana, V. and Ruiz-Hidalgo, J. (2023). Looking behind occlusions: A study on amodal segmentation for robust on-tree apple fruit size estimation. *Comput. Electron. Agric.* 209, 107854.

Gerendás, J. and Ratcliffe, R. G. (2000). Intracellular pH regulation in maize root tips exposed to ammonium at high external pH. *J. Exp. Bot.* 51(343), 207–219.

Goffinet, M. C. and Maloney, K. (1987). histological studies of the development fruit of apple fruit (Malus domestica). *Am. J. Bot.* 74, 629.

Goffinet, M. C., Robinson, T. L. and Lakso, A. N. (1995). A comparison of "Empire" apple fruit size and anatomy in unthinned and hand-thinned trees. *J. Hortic. Sci.* 70(3), 375–387.

Goudriaan, J. and Monteith, J. L. (1990). A mathematical function for crop growth based on light interception and leaf area expansion. *Ann. Bot.* 66(6), 695–701.

Grissom, C. B., Canellas, P. F. and Wedding, R. T. (1983). Allosteric regulation of the NAD malic enzyme from cauliflower: activation by fumarate and coenzyme A. *Arch. Biochem. Biophys.* 220(1), 133–144.

Hamadziripi, E. T., Theron, K. I., Muller, M. and Steyn, W. J. (2014). Apple compositional and peel color differences resulting from canopy microclimate affect consumer preference for eating quality and appearance. *Hort.Sci.* 49(3), 384–392.

Harada, T., Kurahashi, W., Yanai, M., Wakasa, Y. and Satoh, T. (2005). Involvement of cell proliferation and cell enlargement in increasing the fruit size of Malus species. *Sci. Hort.* 105(4), 447–456.

Harker, F. R. and Ferguson, I. B. (1988). Transport of calcium across cuticles isolated from apple fruit. *Sci. Hortic.* 36(3–4), 205–217.

Henton, S. M., Piller, G. J. and Gandar, P. W. (1999). A fruit growth model dependent on both carbon supply and inherent fruit characteristics. *Ann. Bot.* 83(5), 509–514.

Herremans, E., Verboven, P., Hertog, M. L. A. T. M., Cantre, D., van Dael, M., de Schryver, T., van Hoorebeke, L. and Nicolaï, B. M. (2015a). Spatial development of transport structures in apple (Malus × domestica Borkh.) fruit. *Front. Plant Sci.* 6, 679.

Herremans, E., Verboven, P., Verlinden, B. E., Cantre, D., Abera, M., Wevers, M. and Nicolaï, B. M. (2015b). Automatic analysis of the 3-D microstructure of fruit parenchyma tissue using X-ray micro-CT explains differences in aeration. *BMC Plant Biol.* 15, 264.
Hudina, M. and Štampar, F. (2002). Influence of leaf area on the sugar and organic acids content in pear (Pyrus communis) fruits cultivar Williams. *Acta Hortic.* 596, 749–752.
Hulme, A. C., Jones, J. D. and Wooltorton, L. S. C. (1963). The respiration climacteric in apple fruits. *Proc. R. Soc. Lond. B* 158, 514–535.
Jackson, J. E. and Palmer, J. W. (1972). Interception of light by model hedgerow orchards in relation to latitude, time of year, hedgerow configuration and orientation. *J. Appl. Ecol.* 9, 341–357. https://doi.org/10.2307/2402436.
Jackson, J. E. and Sharples, R. O. (1971). The influence of shade and within-tree position on apple fruit size, colour and storage quality. *J. Hortic. Sci.* 46(3), 277–287.
Jackson, J. E., Palmer, J. W., Perring, M. A. and Sharples, R. O. (1977). Effects of shade on the growth and cropping of apple trees. III. Effects on fruit growth, chemical composition and quality at harvest and after storage. *J. Hortic. Sci.* 52(2), 267–282.
Jajo, A., Rahim, M. A., Serra, S., Gagliardi, F., Jajo, N. K., Musacchi, S., Costa, G., Bonghi, C. and Trainotti, L. (2014). Impact of tree training system, branch type and position in the canopy on the ripening homogeneity of abbé fétel' pear fruit. *Tree Genet. Genomes* 10(5), 1477–1488.
Jiang, X. and He, L. (2021). Investigation of effective irrigation strategies for high-density apple orchards in Pennsylvania. *Agronomy* 11(4).
Jing, S. and Malladi, A. (2020). Higher growth of the apple (Malus × domestica Borkh.) fruit cortex is supported by resource intensive metabolism during early development. *BMC Plant Biol.* 20(1), 75.
Johnston, J. W., Hewett, E. W. and Hertog, M. L. A. T. M. (2002). Postharvest softening of apple (Malus domestica) fruit: a review. *N. Z. J. Crop Hortic. Sci.* 30(3), 145–160.
Jones, H. G. and Higgs, K. H. (1982). Surface conductance and water balance of developing apple (Malus pumila mill.) fruits. *J. Exp. Bot.* 33(1), 67–77.
Kader, A. A. (2008). Flavor quality of fruits and vegetables. *J. Sci. Food Agric.* 88(11), 1863–1868.
Kalcsits, L., Musacchi, S., Layne, D. R., Schmidt, T., Mupambi, G., Serra, S., Mendoza, M., Asteggiano, L., Jarolmasjed, S., Sankaran, S., Khot, L. R. and Espinoza, C. Z. (2017). Above and below-ground environmental changes associated with the use of photoselective protective netting to reduce sunburn in apple. *Agric. For. Meteorol.* 237–238, 9–17.
Kappel, F. and Neilsen, G. H. (1994). Relationship between light microclimate, fruit growth, fruit quality, specific leaf weight and N and P content of spur leaves of "Bartlett" and "Anjou" pear. *Sci. Hortic.* 59(3–4), 187–196.
Karim, S. K. A., Allan, A. C., Schaffer, R. J. and David, K. M. (2022). Cell division controls final fruit size in three apple (Malus x domestica) cultivars. *Horticulturae* 8(7), 1–18.
Khan, A. A. and Vincent, J. F. V. (1990). Anisotropy of apple parenchyma. *J. Sci. Food Agric.* 52(4), 455–466.
Kingston, C. M. (1992). Maturity indices for apple and pear. *Hortic. Rev.* 13, 402.
Kramer, P. J. (1983). Problems in water relations of plants and cells. In: International Review of Cytology-a Survey of Cell Biology. Academic Press, 85, 253–286.
Lakso, A. N. and Johnson, R. S. (1990). A simplified dry matter production model for apple using automatic programming simulation software. *Acta Hortic.* 276, 141–148.

Lakso, A. N., Robinson, T. L. and Pool, R. M. (1989). Canopy microclimate effects on patterns of fruiting and fruit development in apples and grapes. In: Wright, C. J. (ed.), *Manipulation of fruiting*, pp. 263–274. 47th Nottingham Easter School, Butterworths, London.

Lakso, A. N., Grappadelli, L. C., Barnard, J. and Goffinet, M. C. (1995). An expolinear model of the growth pattern of the apple fruit. *J. Hortic. Sci.* 70(3), 389–394.

Lakso, A. N., Santiago, M. and Stroock, A. D. (2022). Monitoring stem water potential with an embedded microtensiometer to inform irrigation scheduling in fruit crops. *Horticulturae* 8(12), 1207.

Lang, A. (1990). Xylem, phloem and transpiration flows in developing apple fruits. *J. Exp. Bot.* 41(6), 645–651.

Lapsley, K. G., Escher, F. E. and Hoehn, E. (1992). The cellular structure of selected apple varieties. *Food Struct.* 11, 339–349.

Lau, O. L. (1985). Harvest indices for BC apples. *BC Orchard* 7, 1A–20A.

Lezzer, P., Tustin, S., Corelli-Grappadelli, L., Serra, S., Anthony, B., Dorigoni, A. and Musacchi, S. (2022). Influences of propagation method, rootstock, number of axes, and cultivation site on 'Fuji' scions grown as single or multi-leader trees in the nursery. *Agronomy* 12(1), 224.

Lindo-García, V., Larrigaudière, C., Echeverría, G., Murayama, H., Soria, Y. and Giné-Bordonaba, J. (2019). New insights on the ripening pattern of 'Blanquilla' pears : A comparison between on- and off-tree ripened fruit. *Postharvest Biol. Technol.* 150, 112–121.

Link, H. (2000). Significance of flower and fruit thinning on fruit quality. *Plant Growth Regul.* 31(1/2), 17–26.

Longchamps, L., Tisseyre, B., Taylor, J., Sagoo, L., Momin, A., Fountas, S., Manfrini, L., Ampatzidis, Y., Schueller, J. K. and Khosla, R. (2022). Yield sensing technologies for perennial and annual horticultural crops: a review. *Precis. Agric.* 23(6), 2407–2448.

Lopez, G., Behboudian, M. H., Girona, J. and Marsal, J. (2013). Responses of "conference" pear to deficit irrigation: water relations, leaf discrimination against 13CO2, tree starch content, growth, and recovery after rewatering. *J. Plant Growth Regul.* 32(2), 273–280.

Lopez, G., Boini, A., Manfrini, L., Torres-Ruiz, J. M., Pierpaoli, E., Zibordi, M., Losciale, P., Morandi, B. and Corelli-Grappadelli, L. (2018). Effect of shading and water stress on light interception, physiology and yield of apple trees. *Agric. Water Manag.* 210, 140–148.

Losciale, P., Zibordi, M., Manfrini, L., Morandi, B. and Corelli-Grappadelli, L. (2011). Effect of moderate light reduction on absorbed energy management, water use, photoprotection and photo-damage in peach. *Acta Hortic.* 907, 169–174.

Luo, W. W., Gao, C. X., Zhang, D., Han, M. Y., Zhao, C. P. and Liu, H. K. (2014). [Effects of environmental factors at different altitudes on leaves and fruit quality of Fuji apple]. *Ying Yong Sheng Tai Xue Bao* 25(8), 2243–2250.

Manfrini, L. (2004). Study of cytokinesis activity to ecophysiological factors. Master thesis in agricultural sciences. University of Bologna.

Manfrini, L., Pierpaoli, E., Taylor, J. A., Morandi, B., Losciale, P., Zibordi, M., Corelli Grappadelli, L. and Bastías, R. M. (2012). Precision fruit growing: how to collect and interpret data on seasonal variation in apple orchards. *Acta Hortic.* (932), 461–469.

Manfrini, L., Pierpaoli, E., Zibordi, M., Morandi, B., Muzzi, E., Losciale, P. and Corelli Grappadelli, L. (2015). Monitoring strategies for precise production of high quality fruit and yield in apple in Emilia- Romagna. *Chem. Eng. Trans.* 44, 301–306.

Manfrini, L., Zibordi, M., Pierpaoli, E., Losciale, P., Morandi, B. and Corelli Grappadelli, L. (2019). Development of Precision apple fruit growing techniques: monitoring strategies for yield and high-quality fruit production. *Acta Hortic.* 1261, 191–198.

Mattheis, J. P. and Fellman, J. K. (1999). Preharvest factors influencing flavor of fresh fruit and vegetables. *Postharvest Biol. Technol.* 15(3), 227–232.

McArtney, S., White, M., Latter, I. and Campbell, J. (2004). Individual and combined effects of shading and thinning chemicals on abscission and dry-matter accumulation of "Royal Gala" apple fruit. *J. Hortic. Sci. Biotechnol.* 79(3), 441–448.

McAtee, P. A., Hallett, I. C., Johnston, J. W. and Schaffer, R. J. (2009). A rapid method of fruit cell isolation for cell size and shape measurements. *Plant Methods* 5, 5.

McGlone, V. A., Jordan, R. B., Seelye, R. and Clark, C. J. (2003). Dry-matter: a better predictor of the post-storage soluble solids in apples? *Postharvest Biol. Technol.* 28(3), 431–435.

Mebatsion, H. K., Verboven, P., Ho, Q. T., Mendoza, F., Verlinden, B. E., Nguyen, T. A. and Nicolaï, B. M. (2006a). Modelling fruit microstructure. *Commun. Agric. Appl. Biol. Sci.* 71(1), 203–207.

Mebatsion, H. K., Verboven, P., Ho, Q. T., Mendoza, F., Verlinden, B. E., Nguyen, T. A. and Nicolaï, B. M. (2006b). Modelling fruit microstructure using novel ellipse tessellation algorithm. *C. Comput. Model. Eng. Sci.* 14, 1–14.

Mendoza, F., Verboven, P., Ho, Q. T., Kerckhofs, G., Wevers, M. and Nicolaï, B. (2010). Multifractal properties of pore-size distribution in apple tissue using X-ray imaging. *J. Food Eng.* 99(2), 206–215.

Mendoza, F., Verboven, P., Mebatsion, H. K., Kerckhofs, G., Wevers, M. and Nicolaï, B. (2007). Three-dimensional pore space quantification of apple tissue using X-ray computed microtomography. *Planta* 226(3), 559–570.

Miller, S. A., Smith, G. S., Boldingh, H. L. and Johansson, A. (1998). Effects of water stress on fruit quality attributes of kiwifruit. *Ann. Bot.* 81(1), 73–81.

Morandi, B., Losciale, P., Manfrini, L., Zibordi, M., Anconelli, S., Galli, F., Pierpaoli, E. and Corelli Grappadelli, L. (2014). Increasing water stress negatively affects pear fruit growth by reducing first its xylem and then its phloem inflow. *J. Plant Physiol.* 171(16), 1500–1509.

Morandi, B., Manfrini, L., Zibordi, M., Corelli-Grappadelli, L. and Losciale, P. (2016). From fruit anatomical features to fruit growth strategy: is there a relationship? *Acta Hortic.* 1130, 185–192.

Morandi, B., Manfrini, L., Zibordi, M., Noferini, M., Fiori, G. and Corelli Grappadelli, L. C. (2007). A low-cost device for accurate and continuous measurements of fruit diameter. *Hortscience* 42(6), 1380–1382.

Morandi, B., Zibordi, M., Losciale, P., Manfrini, L., Pierpaoli, E. and Corelli Grappadelli, L. C. (2011). Shading decreases the growth rate of young apple fruit by reducing their phloem import. *Sci. Hortic.* 127(3), 347–352.

Musacchi, S. and Serra, S. (2018). Apple fruit quality: overview on pre-harvest factors. *Sci. Hortic.* 234, 409–430.

Musacchi, S., Iglesias, I. and Neri, D. (2021). Training systems and sustainable orchard management for European pear (Pyrus communis L.) in the Mediterranean area: a review. *Agronomy* 11(9), 1765.

Naor, A., Naschitz, S., Peres, M. and Gal, Y. (2008). Responses of apple fruit size to tree water status and crop load. *Tree Physiol.* 28(8), 1255–1261.

Osorio, S. and Fernie, A. R. (2014). Fruit ripening: primary metabolism. In: Nath, P., Bouzayen, M., Pech, J. C. and Mattoo, A. K. (eds). *Fruit Ripening: Physiology, Signal and Genomics*. Wallingford: CABI Publishing, 15–27.

Palmer, J. W., Harker, F. R., Tustin, D. S. and Johnston, J. (2010). Fruit dry matter concentration: A new quality metric for apples. *J. Sci. Food Agric.* 90(15), 2586–2594.

Palmer, J. W., Wünsche, J. N., Meland, M. and Hann, A. (2002). Annual dry-matter production by three apple cultivars at four within-row spacings in New Zealand. *J. Hortic. Sci. Biotechnol.* 77(6), 712–717.

Patrick, J. W. (1997). PHLOEM UNLOADING: sieve element unloading and post-sieve element transport. *Annu. Rev. Plant Physiol. Plant Mol. Biol.* 48, 191–222.

Pedreschi, R., Franck, C., Lammertyn, J., Erban, A., Kopka, J., Hertog, M., Verlinden, B. and Nicolaï, B. (2009). Metabolic profiling of "Conference" pears under low oxygen stress. *Postharvest Biol. Technol.* 51(2), 123–130.

Policarpo, M., Talluto, G. and Lo Bianco, R. (2006). Vegetative and productive responses of "Conference" and "Williams" pear trees planted at different in-row spacings. *Sci. Hortic.* 109, 322–331.

Ramokonyane, T. M., Sciences, H. and Van Der Merwe, J. A. (2016). Effects of dynamic controlled atmosphere and initial low oxygen stress on superficial scald of 'Granny Smith' apples and 'Packham' s Triumph' pears.

Rapparini, F., Gatti, E., Predieri, S. and Cavicchi, L. (2008). Effect of pear production system on volatile aroma constituents of fruits. *Acta Hortic.* 800(800), 1061–1068.

Rembiałkowska, E. (2007). Quality of plant products from organic agriculture. *J. Sci. Food Agric.* 87(15), 2757–2762.

Reyes, F., DeJong, T., Franceschi, P., Tagliavini, M. and Gianelle, D. (2016). Maximum growth potential and periods of resource limitation in apple tree. *Front. Plant Sci.* 7, 233.

Reyes, F., Pallas, B., Pradal, C., Vaggi, F., Zanotelli, D., Tagliavini, M., Gianelle, D. and Costes, E. (2020). MuSCA: A multi-scale source-sink carbon allocation model to explore carbon allocation in plants. An application to static apple tree structures. *Ann. Bot.* 126(4), 571–585.

Rossi, F., Manfrini, L., Venturi, M., Corelli Grappadelli, L. C. and Morandi, B. (2022). Fruit transpiration drives interspecific variability in fruit growth strategies. *Hortic. Res.* 9, 1–10.

Ruess, F. and Stösser, R. (1993). Untersuchungen über das interzellularsystem bei Apfelfrüchten mit methoden der Digitalen Bildverarbeitung. *Gartenbauwissenschaft* 58, 197–205.

Rutkowski, K. P., Michalczuk, B. and Konopacki, P. (2008). Nondestructive determination of 'Golden Delicious' apple quality and harvest maturity. *J. Fruit Ornam Plant Res.* 16, 39–52.

Sahu, N., Saini, A., Behera, S. K., Sayama, T., Sahu, L., Van Nguyen, V. T. and Takara, K. (2020). Why apple orchards are shifting to the higher altitudes of the Himalayas? *PLoS ONE* 15(7), e0235041.

Saquet, A. A. (2019). Storage of pears. *Sci. Hortic.* 246, 1009–1016.

Saquet, A. A., Streif, J. and Almeida, D. P. F. (2019). Mineral composition and distribution within 'Rocha' pear in relation to internal storage disorders. *Postharvest Biol. Technol.* 158, 111002.

Saudreau, M., Marquier, A., Adam, B. and Sinoquet, H. (2011). Modelling fruit-temperature dynamics within apple tree crowns using virtual plants. *Ann. Bot.* 108(6), 1111–1120.

Scalisi, A., Bresilla, K. and Grilo, F. S. (2017). Continuous determination of fruit tree water-status by plant-based sensors. *Italus Hortus* 24, 39–50.

Serra, S., Leisso, R., Giordani, L., Kalcsits, L. and Musacchi, S. (2016). Crop load influences fruit quality, nutritional balance, and return bloom in 'Honeycrisp' apple. *Hortscience* 51(3), 236–244.

Smith, W. H. (1950). Cell-multiplication and cell enlargement in the development of the flesh of the apple fruit. *Ann. Bot.* 14(1), 23–38.

Sousa, M. L., Gonçalves, M., Fialho, D., Ramos, A., Lopes, J. P., Oliveira, C. M. and De Melo-Abreu, J. P. (2022). Apple and pear model for optimal production and fruit grade in a changing environment. *Horticulturae* 8, 1–25.

Stanley, C. J., Tustin, D. S., Lupton, G. B., Mcartney, S., Cashmore, W. M. and De Silva, H. N. D. (2000). Towards understanding the role of temperature in apple fruit growth responses in three geographical regions within New Zealand. *J. Hortic. Sci. Biotechnol.* 75(4), 413–422.

Stöckle, C. O., Marsal, J. and Villar, J. M. (2011). Impact of climate change on irrigated tree fruit production. *Acta Hortic.* 889, 41–52.

Suni, M., Nyman, M., Eriksson, N. A., Björk, L. and Björck, I. (2000). Carbohydrate composition and content of organic acids in fresh and stored apples. *J. Sci. Food Agric.* 80(10), 1538–1544.

Tombesi, A., Antognozzi, E. and Palliotti, A. (1993). Influence of light exposure on characteristics and storage life of kiwifruit. *N. Z. J. Crop Hortic. Sci.* 21, 87–92.

Torres-Ruiz, J. M., Perulli, G. D., Manfrini, L., Zibordi, M., Lopez Velasco, G., Anconelli, S., Pierpaoli, E., Corelli-Grappadelli, L. and Morandi, B. (2016). Time of irrigation affects vine water relations and the daily patterns of leaf gas exchanges and vascular flows to kiwifruit (Actinidia deliciosa Chev.). *Agric. Water Manag.* 166, 101–110.

Tosin, R., Martins, R., Pôças, I. and Cunha, M. (2022). Canopy VIS-NIR spectroscopy and self-learning artificial intelligence for a generalised model of predawn leaf water potential in Vitis vinifera. *Biosyst. Eng.* 219, 235–258.

Tozzi, F., van Hooijdonk, B. M., Tustin, D. S., Corelli Grappadelli, L., Morandi, B., Losciale, P. and Manfrini, L. (2018). Photosynthetic performance and vegetative growth in a new red leaf pear: comparison of scion genotypes using a complex, grafted-plant system. *Front. Plant Sci.* 9, 404.

Travers, S., Bertelsen, M. G., Petersen, K. K. and Kucheryavskiy, S. V. (2014a). Predicting pear (cv. Clara Frijs) dry matter and soluble solids content with near infrared spectroscopy. *LWT Food Sci. Technol.* 59(2), 1107–1113.

Travers, S., Bertelsen, M. G. and Kucheryavskiy, S. V. (2014b). Predicting apple (cv. Elshof) postharvest dry matter and soluble solids content with near infrared spectroscopy. *J. Sci. Food Agric.* 94(5), 955–962.

Tsoulias, N., Jörissen, S. and Nüchter, A. (2022). An approach for monitoring temperature on fruit surface by means of thermal point cloud. *MethodsX* 9, 101712.

Tustin, D. S. and Van Hooijdonk, B. M. (2016). Can light interception of intensive apple and pear orchard systems be increased with new approaches to tree design? *Acta Hortic.* 1130, 139–144.

Tustin, D. S., Breen, K. C., Hooijdonk, B. M. Van and Words, K. (2022). Scientia Horticulturae Light utilisation, leaf canopy properties and fruiting responses of narrow-row, planar cordon apple orchard planting systems: a study of the productivity of apple. *Sci. Hortic.* 294, 110778.

Tustin, D. S., Hooijdonk, B. M. Van and Breen, K. C. (2018). The Planar Cordon: new planting systems concepts to improve light utilisation and physiological function to increase apple orchard yield potential. *Acta Hortic.* (1228), 1–12.

Verboven, P., Kerckhofs, G., Mebatsion, H. K., Ho, Q. T., Temst, K., Wevers, M., Cloetens, P. and Nicolaï, B. M. (2008). Three-dimensional gas exchange pathways in pome fruit characterized by synchrotron X-ray computed tomography. *Plant Physiol.* 147(2), 518–527.

Volz, R. K., Harker, F. R., Lang, A. and Hallett, I. C. (2002). Texture development in apple fruit: a biophysical perspective. *Progr./Int. Hortic. Congr. Exhib.* XXVIth, 366–367.

Warrington, I. J., Fulton, T. A., Halligan, E. A. and de Silva, H. N. (1999). Apple Fruit Growth and Maturity are Affected by Early Season Temperatures. *J. Am. Soc. Hortic. Sci.* 124(5), 468–477.

Watkins, C. B., Reid, M. S., Harman, J. E. and Padfield, C. A. S. (1982). Starch iodine pattern as a maturity index for granny smith apples: 2. Differences between districts and relationship to storage disorders and yield. *J. Agric. Res.* 25(4), 587–592.

Wei, X., Liu, F., Chen, C., Ma, F. and Li, M. (2014). The Malus domestica sugar transporter gene family: identifications based on genome and expression profiling related to the accumulation of fruit sugars. *Front. Plant Sci.* 5, 569.

Westwood, M. N. (1978). *Temperate-Zone Pomology*. San Francisco: W. H. Freeman and Company.

Westwood, M. N., Batjer, L. P. and Billingsley, H. D. (1967). Cell size, cell number, and fruit density of apples as related to fruit size, position in cluster, and thinning method. *Proc. Amer. Soc. Hort. Sci.* 91, 51–62.

Wünsche, J. and Lakso, A. N. (2000). Apple tree physiology: implications for orchard and tree management. *Compact Fruit Tree* 33, 82–88.

Wünsche, J., Lakso, A., Lenz, F. and Denning, S. (1996). Basis of apple productivity in production. *J. Am. Soc. Hortic. Sci.* 121, 886–893.

Zhang, Q. (2017). *Automation in Tree Fruit Production: Principles and Practice*. CAB International.

Zhang, J., Serra, S., Leisso, S. and Musacchi, S. (2016). Effect of light microclimate on the quality of "d'anjou" pears in mature open-centre tree architecture. *Biosyst. Eng.* 141, 1–11.

Zhang, L. Y., Peng, Y. B., Pelleschi-Travier, S., Fan, Y., Lu, Y. F., Lu, Y. M., Gao, X. P., Shen, Y. Y., Delrot, S. and Zhang, D. P. (2004). Evidence for apoplasmic phloem unloading in developing apple fruit. *Plant Physiol.* 135(1), 574–586.

Zheng, P., Bai, M., Chen, Y., Liu, P. W., Gao, L., Liang, S. J. and Wu, H. (2014). Programmed cell death of secretory cavity cells of citrus fruits is associated with ca2+ accumulation in the nucleus. *Trees* 28(4), 1137–1144.

Zude-Sasse, M., Truppel, I. and Herold, B. (2002). An approach to non-destructive apple fruit chlorophyll determination. *Postharvest Biol. Technol.* 25(2), 123–133.

Chapter 7

Advances in pre-harvest management of apple quality

J. A. Cline, University of Guelph, Canada

1 Introduction

Pre-harvest management of apple orchards has a long history, with reviews previously published on this subject (Musacchi and Serra, 2018). Many textbooks concerned with the physiology and commercial production of temperate tree fruits, including apple, have chapters elucidating key principles to achieve optimum fruit quality (Sansavini et al., 2019).

Fruit quality can be defined in a number of ways, and in itself is a non-specific term that requires definition, depending on the perspective. Notwithstanding, fruit quality is influenced by the orchard environment, especially light, temperature, and water. Fruit appearance is the primary sensory characteristic that consumers consider when evaluating apple fruit quality and making purchase decisions (Canadian Horticultural Council, 1979; Hampson et al., 2002; Nicolai et al., 2007; Jaeger et al., 2016). External appearance includes surface blush and ground colour, fruit that is free of visible defects, and fruit size and shape (Fig. 1). Once purchased, internal textural and organoleptic properties such as fruit firmness, crispness, juiciness, flavour, sweetness (e.g. soluble solids concentration), acidity, and flavour (Harker et al., 2008; Jaeger et al., 1998) and internal defects such as bruising and physiological disorders (Jaeger et al., 2016) become

http://dx.doi.org/10.19103/AS.2023.0127.09

Internal Quality Attributes

Textural properties
- fruit firmness
- crispness

Organoleptic properties
- juiciness
- flavour
- sweetness
- tartness/acidity
- astringency

External Quality Attributes

- surface blush colour
- ground colour
- fruit free of defects caused by insects, rot, misshapen fruit, cracked fruit, surface russet, bitterpit, bruising
- fruit size
- fruit shape

Figure 1 Common external and internal fruit quality attributes of apples.

important. Many of these quality determinants are directly influenced by the type of storage (controlled atmosphere vs. regular atmosphere), stage of maturity when harvested, and how the fruits are handled prior to storage (Watkins, 2017). Musacchi and Serra (2018) provide a comprehensive review of internal fruit qualities influenced by pre-harvest environmental and cultural practices, illustrating several instruments currently available for their assessment.

This chapter will discuss orchard management and environmental conditions that influence fruit quality, excluding those that pertain to disease and pest control strategies, which can also significantly affect several external and internal qualities of apples. Cultivars differ in their inherent quality characteristics and need to be managed accordingly; some cultivars are easier to manage than others. For example, despite its wide consumer appeal and popularity in North America, Honeycrisp is significantly more problematic to grow and requires increased knowledge and more intensive management than other cultivars, such as Gala, to achieve the production of premium quality fruit (Cline, unpublished data). Therefore, pre-harvest orchard management practices are underpinned by a comprehensive understanding of the cultivar being grown, its environment, and tree physiology.

2 Emerging issues affecting pre-harvest management of apple quality

Factors affecting fruit quality can be grouped into environmental factors, such as light and temperature, and those which can be managed horticulturally by manipulating aspects of the orchard, such as the tree canopy, or altering tree nutrition and irrigation management practices. For some fruit quality attributes, plant bioregulators can be used to improve quality attributes and reduce losses at harvest (Fig. 2).

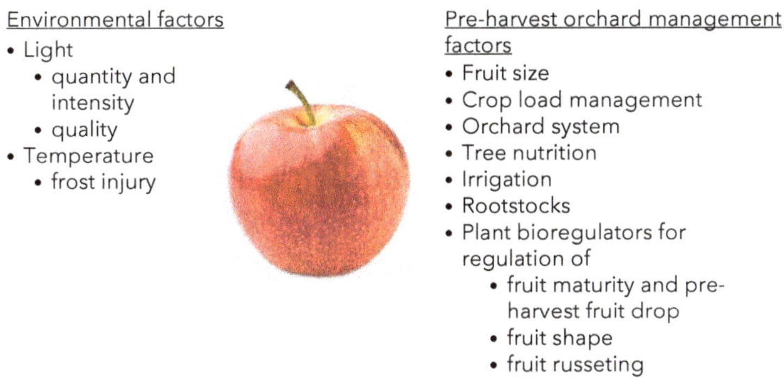

Figure 2 Environmental and pre-harvest orchard management factors affecting fruit quality of apples.

2.1 Fruit colour

Consumers associate red apples with ripeness and good quality, and consequently, red apple cultivars often have better marketability and higher economic value (Meyers and Weidenhamer, 1966; King and Cliff, 2002; Wang et al., 2018a). Further, apples with red skin contain anthocyanins (Allan et al., 2008) which are a class of phenolic compounds with antioxidant properties protecting from chronic diseases such as cancer, heart disease, joint pain, inflammation, diabetes, and cognitive decline (Hyson, 2011). Fruit colour is also a primary determinant in the marketability of fruit. Government regulators set minimum colour-grade standards based on colour intensity and percentage of the fruit skin with blush to ensure poor coloured fruit are excluded (Canadian Food Inspection Agency, 2021). Consequently, fruit colour development is a major factor influencing profitability for apple producers. Presently, several pre-harvest and environmental factors influencing colour development can be optimized, regulated, and sometimes controlled through cultural methods. In addition, apple breeders and researchers are seeking apples with redder skin and red fruit flesh to improve the marketability and health benefits of apples (Chen et al., 2021).

Fruit skin colour is controlled by the biosynthetic pathway of anthocyanins, which is regulated by genetics as well as environmental cues such as sunlight and temperature. The structural genes involved in this pathway in apple have been reviewed by Chen et al. (2021). For bi-coloured apples, chlorophyll in the fruit skin provides the background green colour of fruit (Blanke and Notton, 1992; Jacob-Wilk et al., 1999). During fruit development, a shift in the balance of chlorophyll breakdown and anthocyanin synthesis occurs (Matile et al., 1999), which gradually diminishes the chlorophyll content in mature apples (Jacob-Wilk et al., 1999), leading red or bi-colour apples to take on a red-blush appearance.

2.1.1 The impact of light on fruit colour

Apart from cultivar choice (genetics), light is one of the most important factors that affects colour development in apple. Preceding harvest, apple producers prefer days with high-intensity sunlight and nights with cool temperatures. In China and Japan, fruits are enclosed in light-excluding bags during the growing season and removed just prior to harvest. This practice inhibits anthocyanin accumulation in the peel (Chen et al., 2012) and results in fruits with a pink blush; this is a practical example of the strong association between fruit colour and light.

Takos et al. (2005) determined that light induces the expression of MdMYB1 transcription factor, and others have added to the body of evidence by reporting several MYB transcription factors that exist in anthocyanin biosynthesis (Chen et al., 2021). Light quality and intensity affect anthocyanin development. Ultraviolet and blue light are more effective in promoting colour pigmentation (Mol et al., 1996). Scattering of light is inversely proportional to the fourth power of the wavelength of the ray (Rayleigh's law), so the shorter blue and ultraviolet wavelengths are scattered more than larger wavelengths of light. Diffuse sky radiation favours colour development and explains why atmospheric conditions and latitude have a profound impact on fruit colour and quality. Light intensity can vary due to local environmental conditions such as shading of fruit by the leaf canopy. When light is attenuated due to clouds or smoke, overall solar radiation levels can be reduced by more than 50%, thereby impacting fruit colour. In recent events in Western Canada and the Pacific North-West, many horticulture crops, including apples, have been directly impacted by smoke caused by wildfires (Anonymous, 2020) and high extreme temperatures (Courtney and Mullinax, 2022). This has led to reduced tree productivity and fruit loss from sunburn that causes browning of the fruit skin.

A cultural practice widely adopted for high-value apple cultivars such as Fuji, Gala, and Honeycrisp is the use of reflective mulches to enhance fruit colour (Andris and Crisoto, 1996; Mupambi et al., 2021). Plasticized or woven mulch materials are placed on the orchard floor in the row middles a few weeks prior to harvest to reflect incident radiation into the lower portion of the tree canopy, where fruit colour is most likely to be impaired by canopy shading. Increases in the amount of marketable fruit by as much as 30%, and increased harvest efficiency, by way of the reduced number of times fruits need to be picked for bi-colour apple cultivars, have been reported (Gardiner, 2000). When reflective mulches have been used in combination with dormant and summer pruning programmes, further enhancement of colour development has been achieved.

2.1.2 The impact of temperature on fruit colour

Temperature is another environmental factor that affects numerous plant processes - including colour development. Various studies have reported that fruits exposed to lower temperatures benefit from increased colour development (Lin-Wang et al., 2011). In contrast, higher temperatures inhibit or reduce the synthesis and increase the degradation of anthocyanins (Lakso, 1994; Lin-Wang et al., 2011). It is difficult to reduce air temperature in commercial settings, but one approach, based on the principle of evaporative cooling, is to apply irrigation water to the tree canopy using micro-sprinklers. In a Spanish study on Topred Delicious, Iglesias et al. (2002) found that in warm growing seasons, overhead irrigation applied for 2 h daily for 25–30 days preceding harvest improved red fruit colour, anthocyanin concentration within the fruit skin, and fruit size.

2.2 Orchard management practices

2.2.1 Fruit size

Numerous studies have investigated the importance of orchard management and environmental factors that influence fruit growth and development of apple. Goffinet and Lakso (2017) summarized a body of work pertaining to apple fruit development and fruit size. Much of the earlier research in this area focused on stages of cell division, cell number, and cell size during fruit development (Westwood et al., 1967) and modelling of the sigmoidal pattern of fruit growth defining two distinct phases. Phase 1 represents cell division and occurs within 4–6 weeks of flowering. Phase 2 represents the cell expansion that occurs after Phase 1 until harvest. It is important to recognize that fruit size potential is based on the genetic capacity of the apple cultivar. For example, the size potential of some cultivars such as Gala and Empire is inherently smaller than that of other cultivars such as Honeycrisp or Bramley's Seedling (Goffinet et al., 1995). The most effective method to increase fruit size for a particular cultivar is to increase cell division and ensure the tree is free of stress with sufficient moisture and nutrient availability. Increased cell division can be achieved by thinning flowers at bloom, using chemical fruitlet thinners between petal fall and 20 mm fruitlet diameter, or making exogenous foliar applications of plant bioregulators such as cytokinins (Elfving and Cline, 1993; Wismer et al., 1995). In addition, exposure of fruit to higher ambient air temperatures (Warrington et al., 1999) during the early stages of fruit growth and development can increase fruit size and may explain seasonal differences in fruit size to a limited extent.

With respect to size within flower clusters, fruit can differ markedly based on their position within the corymb; the king (central fruit), which blooms first and is slightly advanced in development, is invariably larger than the subordinate lateral axillary fruits that develop within the flower cluster. Timely removal of lateral fruit through application of post-bloom chemical fruitlet thinners or by natural abscission will reduce inter-fruit competition and result in larger fruit at harvest (Goffinet and Lakso, 2017). Management of crop density, also known as crop load, is the most judicious orchard management practice to improve fruit size at harvest.

Light interception by fruiting spurs and bourse shoots also impacts fruit size. Leaf development and early production of carbohydrates via photosynthesis by bourse shoots are vital for fruit set and the production of large fruit (Abbott, 1960; Ferree and Palmer, 1982). Spur quality is also important and can be limited by exposure to low levels of sunlight caused by shading within the tree canopy (Barritt et al., 1991).

2.2.2 Crop load management

The crop or fruit load on apple trees represents the number of fruit per tree and is singularly the most import factor that affects fruit size at harvest. Fruit thinning is a cultural practice that is directed at reducing fruit load that is financially rewarded for producing larger fruit. The advantages of this practice are well known; reducing the number of flowers or fruits increases fruit size and improves fruit appearance and intrinsic quality, all of which lead to a higher crop value (Costa et al., 2018). Apple producers are rewarded financially for larger fruit, and producers are thus motivated to manage crop loads to a level that optimizes returns – which is dependent on the cultivars, and method of marketing. High-quality fruits that are sold in bulk (individual fruits) rather than in bags generate the greatest value and returns to producers.

Apple trees often produce an overabundance of flowers, and when pollinated, fertilized, and set, produce an excessive number of fruits to achieve marketable fruit size (Solomakhin and Blanke, 2010). Fruit crop load can be reduced through a number of methods, of which the most common is chemical thinning. The goal of successful thinning is to reduce fruit number and achieve the target fruit size using minimal labour and without compromising flower development the following year. This is very difficult in practice and remains a significant challenge for producers, in part because of the unpredictability of fruit set and chemical-induced and natural fruitlet abscission. In addition, despite decades of research, the development of solutions to crop load management has been imperfect and slow due to the interplay of many factors that affect fruit set (Williams, 1979; Williams and Edgerton, 1981) as well as a limited understanding of the mechanisms of fruit abscission (Botton et al., 2011; Larson et al., 2022) and fruit development (Malladi, 2020).

Chemical thinning is the primary method used to manage the crop load of apples, but despite over 60 years of research, it remains an unpredictable production practice with large variations in success across growing seasons and cultivars. Strategies for managing crop load include judicious pruning of flower buds prior to bloom, mechanical and chemical thinning at bloom, and chemical fruitlet thinning after fruit set when fruits are between ~5- and ~15-mm diameter, and hand thinning after natural abscission ('June drop') (McClure and Cline, 2015). Various strategies may be deployed to reach the target crop load, and these differ by region and cultivar. Often bloom thinning compounds are applied at bloom, followed by sequential applications of fruitlet thinners at petal fall until fruits reach 15–20 mm diameter. Beyond this point, they become less responsive to the thinning compounds available. Thinning compounds, concentrations, and suggested application times are typically recommended in local production guides (e.g. Ontario Ministry of Agriculture, Food and Rural Affairs, 2021a). Further details of mechanical approaches, chemical thinning compounds, and advances in fruit thinning have been recently reviewed (Costa et al., 2018; Kon and Schupp, 2018).

It has become increasingly apparent that regardless of the approach, crop load reductions need to be completed early, ideally within 30 days after full bloom and before natural abscission, to reduce the high labour costs of hand thinning, to promote flower bud induction and initiation for the following season (Buban and Faust, 1982; Tromp, 2000), minimize alternate bearing (Cline et al., 2018; Wünsche and Ferguson, 2010), and to maximize fruit size at harvest.

2.2.3 Orchard system

One of the most important orchard management practices that impacts fruit quality is the choice of orchard system at planting (Wagenmakers and Callesen, 1995) and how the orchard is configured. The orchard system consists of several components including rootstock, tree spacing (tree density), method of tree support, tree training, and pruning (Barritt and Van Dalfsen, 1992). Significant progress has been made in the last 50 years in understanding the factors related to light interception by orchards. Specific attention has focused on geometry and canopy characteristics such as tree height, width, spacing, orientation, and leaf area density (Lakso, 1994). A transition over the past 50 years from low-density free-standing trees on semi-dwarfing rootstocks to high-density supported trees on dwarfing rootstocks has led to improved production and product quality. Other benefits include improved precocity, economic returns, labour efficiency, and ease of management (Barritt and Van Dalfsen, 1992). The benefits are beyond the scope of discussion in this current chapter since numerous reviews are available that discuss existing research on various types of orchard systems and their performance (Robinson, 2007;

Close and Bound, 2017; Musacchi and Greene, 2017; Lordan et al., 2018; Neri and Sansavini, 2019).

One of the key tenants in managing the orchard system is balancing vegetative growth with fruiting and renewing vegetative growth to maintain productivity and fruit quality as trees mature (Forshey and Elfving, 1989). The need for mechanization and automation to reduce grower dependence on labour availability and the high cost of labour will be the primary drivers for future transition from three-dimensional orchard systems to two-dimensional planar hedgerow orchards with uniform individual tree units (Zhang et al., 2017). As these changes occur, maintaining the production of superior quality fruit through sustainable practices will be paramount for successful orchard management and production of high-quality fruit.

2.2.4 Tree nutrition

The importance of fruit tree nutrition and its association with fruit quality has long been recognized (Boynton and Oberly, 1966; Neilsen and Neilsen, 2003). With perennial crops such as apple, tree nutrition impacts tree growth and fruiting, tree health, and storage potential of fruits. There is a myriad of considerations when applying soil and foliar nutrients to apple trees, and a more complete discussion of these is beyond the scope of this chapter. Marcelle (1995), Fallahi et al. (2010) Tagliavini (2019), and Sharples (1980) are excellent sources that summarize this topic. In addition, local production guidelines (e.g. Penn State University, 2022) represent excellent sources of information on local nutrition guidelines.

Several mineral nutrients can influence fruit quality and disorders of apple. Among these, nitrogen (N), potassium (K), phosphorous (P), calcium (Ca), magnesium (Mg), and boron (B) are most often correlated to apple fruit quality and post-harvest disorders. Leaf and fruit mineral analysis aid in the prediction of fruit quality and how to take corrective action in situations when levels are outside optimal sufficiency ranges. Extensive research has been conducted to predict apple fruit quality in response to soil, fertigation, and foliar nutritional programmes, and various orchard factors and cultural practices - such as irrigation and rootstocks - on apple fruit mineral composition and quality. Evidence-based best practices include conducting leaf analyses to guide early pre-bud break (dormant) applications of N, P, K, and B and applying foliar sprays of N, B, Mg, and micro-nutrients should deficiencies develop during fruit growth and tree development. Bi-weekly prophylactic foliar sprays of Ca (typically calcium chloride) are required to control cultivars prone to bitter pit, a Ca-related fruit disorder (Ferguson and Watkins, 1989; de Freitas et al., 2015). Red Delicious, Honeycrisp, Cortland, and Northern Spy are examples of North American cultivars with a high predisposition to bitter pit, while others, such as

Gala, are rarely affected. Hence, bitter pit incidence is very cultivar dependent, and awareness of local cultivars and industry recommendations is required for implementing optimal control measures.

Relationships between N and fruit quality are complex. Nitrogen is associated with tree vigour and fruit size, but excessive tree N can reduce red colour, firmness and storage potential of fruit, delay fruit maturity, and increase pre-harvest fruit drop (PFD), all contributing to reduced packouts and marketable fruit (Neilsen et al., 1999; Neilsen and Neilsen, 2003). Potassium also influences fruit quality if imbalances exist in both the tree and fruit. Fruit size, colour, and acidity are positively related to K levels (Boynton and Oberly, 1966; Stiles, 1994; Nava et al., 2007). Excessive K levels are associated with increased scald, bitter pit, and breakdown after storage (Weeks et al.,1965). Magnesium deficiency can lead to PFD in cultivars prone to drop prematurely, such as McIntosh and Honeycrisp (Arseneault and Cline, 2018). Foliar spray of magnesium sulphate will correct deficiency in the current year only and should be combined with soil application for a longer-term solution (Ontario Ministry of Agriculture, Food and Rural Affairs, 2021b). Reports concerning the direct relationship between P and pre-harvest fruit quality are conflicting. Differences in the results of various studies appear to be associated with a combination of varietal and climatic differences, as well as other factors.

The cations K, Mg, and Ca interact in the soil and plant tissue, including fruit, leading to changes in fruit quality if imbalances occur (Bramlage et al., 1980). Overall, understanding relationships between pre-harvest tree nutrition, post-harvest fruit quality, and orchard practices informs various management decisions, such as fertilization, pruning, and irrigation.

2.2.5 Irrigation

Fruit growth requires substantial support of carbohydrates, nutrients, and water, and deficits of any of these resources can impact fruit quality. The requirement for careful attention to the water status of the tree particularly should be self-evident. Soil moisture, tree water status, and the benefits of irrigation when natural precipitation is inadequate to support the tree and to optimize yield have been the subject of numerous pomological studies. Many of these were first conducted in arid regions where the risk of water stress is the greatest, such as the Galilee region in northern Israel, and the east of the cascade mountains in the State of Washington, USA, and in the Okanagan Valley, British Columbia, Canada. Orchard modernization and intensification have been accompanied by more recent extremes in precipitation and temperature as a result of climate change, leading irrigation to become ubiquitous in most apple production regions globally. Even when precipitation may be adequate to support apple tree growth in more humid production regions, irrigation is being increasingly utilized to minimize the risk associated with the high establishment costs of new high-density orchards. Risks associated with weather-related crop loss must be managed properly to optimize the return on investment and to ensure financial success; providing supplemental water to maintain fruit quality is one such practice. Naturally, irrigation is only possible with the availability of water and the infrastructure to support it. Approaches to how irrigation water is applied (trickle, micro-sprinklers, overhead), the amount, and scheduling (frequency) have been the subject of many studies and indeed will vary across growing regions – simply because soil texture, water availability, natural rainfall, and the environment vary across production regions. In some regions where water is limited and even restricted for agricultural use (e.g. parts of Australia, California, and Washington State), regulated deficit irrigation (RDI) is practiced to conserve water for times of critical fruit growth and development and to minimize the negative effects of water stress on fruit quality. RDI can also be used to control excess vegetative growth and to cause terminal buds to set earlier (Ebel et al., 1995; Naor, 2006), which may indirectly improve fruit quality.

The economic benefits of ensuring a continuous supply of quality fruit to retail markets are the primary driver for adopting irrigation by apple producers. Apple trees can respond with significant increases in fruit size and yield when water is not limiting; however, the frequency, duration, and variation in soil and environment across trials have led to inconsistent reports on the effects of irrigation on apple tree growth and fruit quality. Furthermore, improvements in fruit quality may not be uniform for all apple cultivars. In commercial orchards, there is a wide diversity of combinations of cultivars, rootstocks, training systems, crop loads, potential fruit sizes, and irrigation efficiencies. Moreover, since each combination can affect irrigation requirements differently, these combinations need to be considered when using irrigation to improve fruit quality and yield.

Future advances in irrigation management for apple will likely focus on improvements in measuring plant water status and developing rootstock genetics to withstand greater drought stress. Recent technological advances have led to the development of micro-tensiometers that attach to the tree trunk above the graft union, which provide continuous stem water potential measurements (Blanco and Kalcsits, 2021). Such technology provides a direct measurement of water stress rather than indirect potential stress through measurement of soil moisture levels, for example. As such, it may be used in the future to schedule irrigation or make the most judicious use of water when it is limited.

Rootstock genotype differences in tree response to water stress have been observed empirically, but there lies some difficulty in understanding the bases of these responses and whether they represent indirect effects, crop load, for example, or direct physiological effects (Lakso, 2003). Rootstocks can affect root architecture, root morphology, hydraulic conductivity, and leaf conductance of the scion. Drought-induced molecular and biochemical changes in rootstocks in response to water status are also involved in drought tolerance. Development of increased drought-tolerant apple rootstocks in the future will reduce the impact of water stress on fruit quality. For greater discussion of this topic, see reviews by Webster and Wertheim (2003), Tworkoski et al. (2016), Marini and Fazio (2018), and Wang et al. (2019).

2.2.6 Rootstocks

The interaction between the scion and rootstock is complex because the root system provides water and nutrients to the scion but requires assimilation from the scion. Marini and Fazio (2018) provided a compressive review of apple rootstocks and their influence on tree physiology. Extensive research on rootstock performance has been conducted for decades in North America under the auspices of the USDA NC-140 technical committee (http://www.nc140.org/). Rootstocks impart a host of attributes on the scion including vigour, precocity, productivity, flower intensity, and tolerance of abiotic and biotic stresses. Trees on dwarfing rootstocks are smaller and typically intercept and distribute more sunlight into the tree canopy, directly affecting both fruit and fruiting spur quality. Fruits that develop in the shade contain less sugar and are less mature compared to those exposed to the full sun (Heinicke, 1964; Zhang et al., 2016). Mature apple trees on dwarfing rootstocks with full crops can partition up to 75% of their total seasonal dry matter into fruit when orchards are optimally managed (Palmer, 1992; Palmer et al., 2002; Wünsche and Ferguson, 2010). Consequently, dwarfing rootstocks shift the balance of growth from vegetative to reproductive which results in earlier fruiting and more productive orchards.

The direct influence of the rootstock on fruit quality is equivocal because of the complex interaction with the scion. The effect of rootstocks on crop load is one of the most important factors in orchard management influencing yield and fruit quality. Rootstock studies on fruit quality often fail to consider the confounding factors of crop load on fruit size and light interception (Autio, 1991; Marini et al., 2002). As a result, fruit quality differences reported among rootstocks may be linked to differences in the light environment in which the fruit is grown, rather than a direct rootstock effect (Barritt et al., 1997). While much research on apple rootstock describes the influence of clonal and seedling rootstocks on tree vigour and productivity, many studies have also investigated rootstocks' effects on fruit characteristics and storage quality (Drake et al., 1988; Autio, 1991; Autio et al., 1996; Barden and Marini, 1992; Barritt et al., 1996). Rootstocks have been reported to influence fruit size, firmness, colour, soluble solids content, carbohydrates and mineral concentration, maturity, ripening, storage quality, and respiration rates at harvest and during storage (Barritt et al., 1997). Results, however, are inconsistent as they vary across locations, seasons, and cultivars (Marini et al., 2002). No particular rootstocks have been identified that improve fruit quality, and rootstock selection is based instead on vigour control, improved productivity, and resistance to abiotic and biotic stress. Nevertheless, the root system of apple trees is responsible for absorbing minerals and nutrients for the plant, and differences in rootstock genotypes and nutrient updates have been a continued area of recent focus.

In general, it appears that rootstocks have little influence on N and K status and have a greater effect on Ca and Mg uptake and distribution, and Ca and Mg are widely associated with bitter pit propensity (Marini and Fazio, 2018). Valverdi and Kalcsits (2021a, 2021b) conducted a study on 2- and 3-year-old high-density planting (3000 trees/ha) with two irrigation regimes in Washington State. Honeycrisp trees grown on G.41, G.890, M.9-T337, and Bud. 9 rootstocks did not affect calcium uptake. However, rootstock-induced vigour and the effect of irrigation were both found to contribute to nutrient imbalances in leaves and fruit that could lead to physiological disorders in Honeycrisp apple. Rootstocks B.9 and M.9 had a lower probability of bitter pit incidence than 'G.41' and 'G.890' in both years.

2.2.7 Plant bioregulators

2.2.7.1 Fruit maturity and pre-harvest fruit drop

Ethylene is one of several plant growth regulators that affects growth and developmental processes including abscission of plant organs, fruit ripening, and senescence. Climacteric fruits, including apples, are characterized by an exponential increase in respiration triggered by autocatalytic ethylene production, thus signalling the onset of ripening. Ethylene can consequently and profoundly influence apple quality, and these effects can be beneficial or deleterious depending on the fruit, its ripening stage, the storage strategy adopted, and its final use. Endogenous ethylene is essential for fruit ripening and maturation and is often mediated through various management practices to improve fruit quality prior to harvest or to slow the rate of fruit quality decline during harvest as apples mature.

Two ethylene-mediated plant bioregulators are widely used in apple production to reduce the negative impacts of ethylene on fruit quality. Aminoethoxyvinylglycine (AVG) is used to reduce PFD and regulate the undesirable effects of ethylene on fruit softening (Byers, 1997; Greene, 2002; Arseneault and Cline, 2016; Arseneault and Cline, 2018). By virtue of interfering with ethylene biosynthesis (Baker et al., 1978; Bangerth, 1978), the ripening process can be slowed if apple fruits are treated with AVG 2–4 weeks prior to harvest to reduce the autocatalytic production of ethylene. AVG can delay apple maturity on the tree, resulting in a longer harvest period. Other potential fruit quality benefits of AVG include increases in soluble solids, colour, fruit size, fruit firmness and reductions in starch hydrolysis, incidence of stem-end cracking, scarf skin, and water core. AVG can reduce the fruit surface colour of some cultivars such as Gala and Honeycrisp. As a result, a reduction in the application rate of AVG may need to be considered when using AVG on colour-sensitive cultivars (Arseneault and Cline, 2018). Numerous studies have evaluated the response of apple trees and other temperate climacteric fruits to AVG, but responses vary based on cultivar, environmental conditions, time of application, and tree stress levels caused by insects or drought.

The ethylene-related bioregulator 1-methylcyclopropene (1-MCP) has been widely adopted by the apple industry to control post-harvest ripening and preserve fruit quality in storage (Watkins et al., 2000). 1-MCP inhibits the perception of ethylene by competing for the ethylene-binding sites in the fruit, thereby slowing fruit ripening, tissue softening, volatile compound production, and fruit breakdown during storage (Blankenship and Dole, 2003; Watkins, 2006, 2008).

When first developed, 1-MCP was marketed as a post-harvest treatment that could be applied to apples in sealed controlled atmosphere storage rooms (Blankenship and Dole, 2003). With the post-harvest success of 1-MCP, interest has grown in the use of a pre-harvest exogenous foliar form of 1-MCP (Harvista™) and its effects on fruit quality. Yuan and Carbaugh (2007) and Yuan and Li (2008) indicate that the sprayable form of 1-MCP reduced the production of ethylene, maintained fruit firmness, and decreased fruit drop in Delicious apple trees compared with untreated trees. Several other studies support the benefits of pre-harvest application of 1-MCP on post-harvest fruit quality. These include the work of Nock et al. (2009), who conducted a study on Delicious, Empire, Gala, Jonagold, Macoun, McIntosh, and Honeycrisp and found that pre-harvest application of 1-MCP markedly delayed PFD in susceptible cultivars, reduced internal fruit ethylene concentrations (IEC) and starch hydrolysis, and increased fruit firmness. However, the degree of benefit differed by cultivar. Elfving et al. (2007) found that Scarletspur Delicious and Cameo apples treated with 1-MCP had less flesh firmness loss and IEC up to 225 days of storage during a 7-day, post-storage, shelf-life simulation at room temperature. Sprayable 1-MCP also reduced starch hydrolysis, IEC, and fruit drop at harvest. Application closer to harvest improved the effect of sprayable 1-MCP on the control of flesh firmness loss and IEC. Treatment with 1-MCP showed little effect on soluble solids concentration, titratable acidity, or skin or flesh colour in 'Delicious' but slightly increased soluble solids concentration and titratable acidity in 'Cameo' apples. In another study on Honeycrisp, Al Shoffe et al. (2021) demonstrated that pre-harvest 1-MCP treatments suppressed Honeycrisp fruit IEC, delayed colour development, reduced soft scald development, and increased fruit firmness after storage, but increased bitter pit development and the incidence of CO_2 injury.

2.2.7.2 Fruit shape

The shape or 'typiness' of select apple cultivars, such as Delicious, is important for consumer recognition. Fruit shape can vary by different climactic regions. Typiness of Delicious is favoured in climates with warm days and cool nights (Westwood and Burkhart, 1966; Greenhalgh and Godley, 1976). Fruit shape is also affected by the pollination of each ovary in the fruit; if pollination is sporadic, a greater proportion of fruit will be mis-shaped, leading to lower fruit quality and decreased packouts. In some cultivars, fruits are asymmetric, with one side of the fruit lacking seeds. Exogenous applications of gibberellic acid (GA_4 and GA_3) and cytokinins are effective for elongating fruit when they are applied at bloom and up to three weeks afterward during Phase 1 of fruit growth (Burak and Büyükyilmaz, 1997). The use of plant bioregulators to increase the typiness and shape of apples is arguably cosmetic. Thus, since fruit shape does not affect internal organoleptic and textural properties, its use is no longer acceptable in an environment where consumers expect fruit to be grown with reduced or no synthetic agrochemicals.

2.2.7.3 Fruit russeting

Fruit russet directly affects fruit appearance and consequently can result in marked reductions in marketable yield and crop loss. Apart from specific apple cultivars, such as Reinette, Egremont Russet, or Golden Russet for example, consumers prefer fruit without surface imperfections. Fruit russeting can also result in post-harvest water loss and fruit shrinkage in post-harvest storage (Macnee et al., 2020; Khanal et al., 2021), reduced shine, increased shrivel, and reduced packed weight (Winkler et al., 2022). The aetiology of fruit russet is borne in both environmental and biological causes. Orchard factors that induce russeting generally occur early during fruit ontogeny when rapid cell division of epidermal cells exceeds expansion of the cuticle, leading to the formation of micro-cracks in the cuticle and a secondary periderm layer (Winkler et al., 2022). This physical injury to the fruit surface is manifested in brown corky blemishes on the fruit epidermis as fruit mature and is particularly noticeable on yellow or green fruit such as Golden Delicious (Faust and Sherer, 1972).

The biotic drivers of russet include expansion-growth-induced strain, surface wetness early during fruit development, temperature, light, relative humidity, mechanical damage, frost injury during bloom, some pests and diseases (e.g. yeasts) (Lindow et al., 1998; Gildemacher et al., 2006), and some agrochemicals, such as copper sprays. Winkler et al. (2022) provide an excellent summary compiled from the literature on the effects of fungicides, surfactants, and foliar fertilizers that have reported an increase or decrease.

Fruit russeting has been associated with fruit shape, orchard location, and endogenous levels of gibberellin (Eccher, 1986; Eccher and Hajnajari, 2006). Compared to fruit grown at lower elevations and warmer regions, fruit grown in cooler environments are more elongated, have thicker cuticles, contain higher levels of endogenous GA, and exhibit a lower incidence of russeting (Eccher, 1986). Applications of GA_{4+7} are the primary method to reduce physiological skin russeting, increase the fruit L/D ratio, and thicken the cuticle (Taylor, 1975; Eccher, 1986). Typically, russet-sensitive cultivars such as Golden Delicious are treated with 15–25 mg GA_{4+7}/L over a period of 28–40 days after bloom (Taylor and Knight, 1986; Elfving and Allan, 1987; Miller, 1988; Greene, 1993; McArtney et al., 2007; BioSciences, 2022). For a comprehensive review of the biochemistry and molecular biology of russet formation, the reader is referred to the excellent recent reviews by Faust and Sherer (1972), Wang et al (2016), Macnee et al. (2020), and Winkler et al. (2022).

3 Conclusion and future trends in research

Cultivar genetics determine the quality attributes of a cultivar, such as its flavour, texture, and colour. How the cultivar is managed in the orchard determines whether this full potential is achieved. Orchard management practices can vary significantly among cultivars. Some are easier to manage in the orchard than others, and this can affect the economic determinants in growing a particular cultivar. Hence, research on pre-harvest factors that enhance fruit quality will likely focus on cultivar-specific solutions. For example, there is very focused research on crop load management and pruning practices for producing Cosmic Crisp™ (WA 38) apples, which is a relatively new apple cultivar grown exclusively upon release in Washington, USA (Musacchi et al., 2017; Anthony et al., 2017). Breeding of new apple genotypes with improved internal fruit qualities, disease resistance, and horticultural attributes such as annual bearing and ease of management will drive innovations in apple fruit quality in the future. Molecular genetics will help to shed light on the underlying mechanisms that give rise to specific fruit quality characteristics, leading to new ways to optimize fruit quality characteristics in pre-harvest and post-harvest storage.

The reduced availability of skilled labour and high labour costs will drive innovation in the orchard industry, especially for apple fruit. Methods to reduce variation in tree size and develop uniform hedgerow canopies will lend themselves to the adaption of technology that will automate processes such as pruning (He and Schupp, 2018) and harvesting. Any new innovation must improve upon existing technology without sacrificing fruit quality and orchard sustainability. In addition to new methods to manage orchards, new rootstock genotypes and cultivars with improved quality attributes will be introduced to meet consumer demand for new products.

Orchard management will become more precise with advances in image system technology and agronomic sensors combined with deep learning algorithms, and some of these technologies are already commercially available. There are platforms that can accurately detect and count apple flower clusters at bloom and fruits per tree to aid in fruit thinning and crop load management. This platform can also determine the effects of light interception, rootstock, and row orientation on flower cluster number, crop load, yield, and tree geometry (Scalisi et al., 2021). Such information will lead to new, yet unrealized, levels of orchard management to make informed evidence-based decisions to improve fruit quality and optimize yield. Advances in weather monitoring coupled with design-aid platforms will provide orchardists with improved forecasting of pest and disease outbreaks leading to more efficient and time-sensitive use of pesticides for their control. Collectively, new technology to simplify and expand the ability to obtain and provide information to the tree fruit industry will result in improved orchard management practices and improved fruit quality, helping to sustain the economic viability of the apple industry.

4 Where to look for further information

Managing pre-harvest factors affecting fruit quality in apples has a long history but has become more important in relation to long-term storage and shipping of fruits through distribution channels. Consumers have become more discerning and particular in their purchasing habits and expect high-quality apples that are fresh, firm, crisp, and juicy even when out of season. There has been constant change in the varieties of apples grown, with breeders and private breeding programmes releasing new apple cultivars with a range of quality attributes and flavours. Increasing global production has forced producers to provide exceptional quality fruit to compete, resulting in increased varieties of apples grown and available in supermarkets and retail markets. Several pomology textbooks concerning tree fruit physiology and commercial production of temperate tree fruits, including apples, have chapters highlighting key principles pertaining to fruit quality (Childers et al., 1995; Tromp et al., 2005). The Horticultural Reviews series contains comprehensive reviews of pertinent research on horticultural crops with information relevant to the fruit quality of apples. Many countries have national and local horticultural societies which hold meetings to update horticultural research including various factors that affect fruit quality. Some produce digital proceedings accessible through the internet. These include the Ontario Fruit and Vegetable Convention (https://www.ofvc.ca/),

New York Fruit Quarterly (https://nyshs.org/fruit-quarterly/), Washington State Tree Fruit Association (https://wstfa.org/annual-meeting/), and International Tree Fruit Association (https://www.ifruittree.org/). The International Society of Horticultural Science (ISHS) (http://www.ishs.org) serves a similar international function with working groups and periodic symposium pertinent to the subject of this chapter. Some symposia of interest include:

- International symposium on flowering, fruit set, and alternate bearing;
- International symposium on integrating canopy, rootstock, and environmental physiology in orchard systems;
- International apple symposium; and
- International symposium on plant bioregulators in fruit production.

There are also several popular press periodicals covering various aspects of apple production (Good fruit grower https://www.goodfruit.com; American Fruit Grower https://www.meistermedia.com/american-fruit-grower/; Fruit Grower News https://fruitgrowersnews.com/; The Orchardist https://www.hortnz.co.nz/news-events-and-media/magazines/; European Fruit Magazine https://fruitmagazine.eu/; The Fruit Grower Magazine https://hortnews.com/). For practical production information, the extension service in many countries has information online and in print. Some examples include: https://orchardmanuals.com.au/, https://onfruit.ca/, http://treefruit.wsu.edu/, https://www.canr.msu.edu/apples/index, and https://extension.psu.edu/forage-and-food-crops/fruit.

5 References

Abbott, D. L. 1960. The bourse shoot as a factor in the growth of apple fruits. *Ann. Appl. Biol.* 48(2):434–438.

Al Shoffe, Y., Nock, J. F., Zhang, Y. and Watkins, C. B. 2021. Physiological disorder development of 'Honeycrisp' apples after pre-and post-harvest 1-methycyclopropene (1-MCP) treatments. *Postharvest Biol. Technol.* 182:111703.

Allan, A. C., Hellens, R. P. and Laing, W. A. 2008. MYB transcription factors that colour our fruit. *Trends Plant Sci.* 13(3):99–102.

Andris, H. and Crisoto, C. H. 1996. Reflective materials enhance 'Fuji' apple color. *Calif. Agric.* 50(5):27–30.

Anonymous. 2020. Wind, smoke may affect Washington's apple crop this year. Available at: https://www.yaktrinews.com/wind-smoke-may-affect-washingtons-apple-crop-this-year/ (accessed June 6, 2022).

Anthony, B. M., Musacchi, S. and Serra, S. 2017. Crop load effect on fruit quality in Cosmic Crisp (TM). *Hort.Sci.* 9(52):S409–S409.

Arseneault, M. H. and Cline, J. A. 2016. A review of apple preharvest fruit drop and practices for horticultural management. *Sci. Hort.* 211:40–52.

Arseneault, M. H. and Cline, J. A. 2018. AVG, NAA, boron, and magnesium influence preharvest fruit drop and fruit quality of 'Honeycrisp' apples. *Can. J. Plant Sci.* 98(3):741–752.

Autio, W. R. 1991. Rootstocks affect ripening and other qualities of 'Delicious' apples. *J. Am. Soc. Hortic. Sci.* 116(3):378–382.

Autio, W. R., Hayden, R. A., Micke, W. C. and Brown, G. R. 1996. Rootstock affects ripening, color, and shape of 'Starkspur Supreme Delicious' apples in the 1984 NC-140 cooperative planting. *Fruit Var. J.* 50:45–53.

Baker, J. E., Lieberman, M. and Anderson, J. D. 1978. Inhibition of ethylene production in fruit slices by a rhizobitoxine analog and free radical scavengers. *Plant Physiol.* 61(6):886–888.

Bangerth, F. 1978. The effect of a substituted amino acid on ethylene biosynthesis, respiration, ripening and preharvest drop of apple fruits. *J. Am. Soc. Hort. Sci.* 103(3):401–404.

Barden, J. A. and Marini, M. E. 1992. Maturity and quality of 'Delicious' apples as influenced by rootstock. *J. Am. Soc. Hortic. Sci.* 117(4):547–550.

Barritt, B. H., Konishi, B. S. and Dilley, M. A. 1997. Tree size, yield and biennial bearing relationships with 40 apple rootstocks and three scion cultivars. *Acta Hortic.* 451:105–112.

Barritt, B. H., Konishi, B. S., Drake, S. R. and Rom, C. R. 1997. Influence of sunlight level and rootstock on apple fruit quality. *Acta Hortic.* 451:569–578.

Barritt, B. H., Rom, C. R., Konishi, B. J. and Dilley, M. A. 1991. Light level influences spur quality and canopy development and light interception influence fruit production in apple. *Hortscience* 26(8):993–999.

Barritt, B. H. and Van Dalfsen, K. B. 1992. *Intensive Orchard Management: A Practical Guide to the Planning, Establishment, and Management of High Density Apple Orchards*. Yakima, WA: Good Fruit Grower, p. 211.

BioSciences, V. 2022. ProVide® 10SG plant growth regulator. Available at: https://s3-us-west-1.amazonaws.com/agrian-cg-fs1-production/pdfs/ProVider_10_SG_Plant_Growth_Regulator_Label2b.pdf (accessed July 27, 2022).

Blanco, V. and Kalcsits, L. 2021. Microtensiometers accurately measure stem water potential in woody perennials. *Plants (Basel)* 10(12):2780.

Blanke, M. M. and Notton, B. A. 1992. Light transmission into apple fruit and leaves. *Sci. Hort.* 51(1–2):43–53.

Blankenship, S. M. and Dole, J. M. 2003. 1-Methylcyclopropene: A review. *Postharvest Biol. Technol.* 28(1):1–25.

Botton, A., Eccher, G., Forcato, C., Ferrarini, A., Begheldo, M., Zermiani, M., Moscatello, S., Battistelli, A., Velasco, R., Ruperti, B. and Ramina, A. 2011. Signaling pathways mediating the induction of apple fruitlet abscission. *Plant Physiol.* 155(1):185–208.

Boynton, D. and Oberly, G. H. 1966. Apple nutrition. In: Childers, N. F. (Ed.), *Nutrition of Fruit Crops*. New Brunswick, NJ: Horticultural Publications, pp. 1–50.

Bramlage, W. J., Drake, M. and Lord, W. J. 1980. Influence of mineral nutrition on the quality and storage performance of pome fruits grown in North America. In: Atkinson, D., Jackson, J. E., Sharples, R. O. and Waller, W. M. (Eds.), *Mineral Nutrition of Fruit Trees*. London, England: Butterworths, pp. 29–39.

Buban, T. and Faust, M. 1982. Flower bud induction in apple trees. *Hort. Rev.* 4:174–203.

Burak, M. and Büyükyilmaz, M. 1998. Effect of Promalin on fruit shape and quality of Starking Delicious apple cultivar. *Acta Hortic.* 463:365–370.

Byers, R. E. 1997. Effects of aminoethoxyvinylglycine (AVG) on preharvest drop, maturity, and cracking of several apple cultivars. *J. Tree Fruit Prod.* 2(1):77–97.

Canadian Food Inspection Agency. 2021. Canadian grade compendium. *Fresh Fruit or Vegetables*. Available at: https://inspection.canada.ca/about-cfia/acts-and-regulations/list-of-acts-and-regulations/documents-incorporated-by-reference/canadian-grade-compendium-volume-2/eng/1519996239002/1519996303947?chap=0 (accessed June 6, 2022).

Canadian Horticultural Council. 1979. *A Survey of Canadian Consumers Concerning Apples and Other Fresh Fruit*. Prepared for the Canadian Horticultural Council by R.R. Research Services. Toronto, ON: Ronalds-Reynolds and Co., p. 264.

Chen, C. S., Zhang, D., Wang, Y. Q., Li, P. M. and Ma, F. W. 2012. Effects of fruit bagging on the contents of phenolic compounds in the peel and flesh of 'Golden Delicious', 'Red Delicious', and 'Royal Gala' apples. *Sci. Hort.* 142:68–73.

Chen, Z., Yu, L., Liu, W., Zhang, J., Wang, N. and Chen, X. 2021. Research progress of fruit color development in apple (*Malus domestica* Borkh.). *Plant Physiol. Biochem.* 162:267–279.

Childers, N. F., Morris, J. R. and Sibbet, G. S. 1995. *Modern Fruit Science*. Gainesville, FL: Horticultural Publications.

Cline, J. A., Bakker, C. J. and Gunter, A. 2019. Response of Royal Gala apples to multiple applications of chemical thinners and the dynamics of fruitlet drop. *Can. J. Plant Sci.* 99(1):1–11.

Close, D. C. and Bound, S. A. 2017. Advances in understanding apple tree growth: Rootstocks and planting systems. In: *Achieving Sustainable Cultivation of Apples*. Cambridge, UK: Burleigh Dodds Science Publishing, pp. 35–52.

Costa, G., Botton, A. and Vizzotto, G. 2018. Fruit thinning: Advances and trends. *Hort. Rev.* 46:185–226.

Courtney, R. and Mullinax, T. J. 2022. Apples in extreme heat. *Goodfruit Grower*. Available at: https://www.goodfruit.com/apples-in-extreme-heat/ (accessed June 6, 2022).

de Freitas, S. T., do Amarante, C. V. T. and Mitcham, E. J. 2015. Mechanisms regulating apple cultivar susceptibility to bitter pit. *Sci. Hort.* 186:54–60.

Drake, S. R., Larsen, F. E., Fellman, J. K. and Higgins, S. S. 1988. Maturity, storage quality, carbohydrate, and mineral content of 'Goldspur' apples as influenced by rootstock. *J. Am. Soc. Hort. Sci.* 113(6):949–952.

Ebel, R. C., Proebsting, E. L. and Evans, R. G. 1995. Deficit irrigation to control vegetative growth in apple and monitoring fruit growth to schedule irrigation. *Hortscience* 30(6):1229–1232.

Eccher, T. 1986. Russeting and shape of 'Golden Delicious' apples as related to endogenous GA content of fruitlets. *Acta Hortic.* 179:769–770.

Eccher, T. and Hajnajari, H. 2006. Fluctuations of endogenous gibberellin A4 and A7 content in apple fruit with different sensitivity to russet. *Acta Hortic.* 727:537–544.

Elfving, D. C. and Allen, O. B. 1987. Effect of gibberellin A4+7 applications on 'Golden Delicious' fruit russet. *Crop Res.* 27:11–18.

Elfving, D. C. and Cline, R. A. 1993. Cytokinin and ethephon affect crop load, shoot growth, and nutrient concentration of Empire apple trees. *Hort. Sci.* 28(10):1011–1014.

Elfving, D. C., Drake, S. R., Reed, A. N. and Visser, D. B. 2007. Preharvest applications of sprayable 1-methylcyclopropene in the orchard for management of apple harvest and postharvest condition. *Hortscience* 42(5):1192–1199.

Fallahi, E., Fallahi, B., Neilsen, G. H., Neilsen, D. and Peryea, F. J. 2010. Effects of mineral nutrition on fruit quality and nutritional disorders in apples. *Acta Hortic.* 868:49–60.

Faust, M. and Sherer, C. B. 1972. Russeting of apples, an interpretive review. *Hort.Sci.* 7(3):233–235.

Ferguson, I. B. and Watkins, C. B. 1989. Bitter pit in apple fruit. *Hort. Rev.* 11:289–355.

Ferree, D. C. and Palmer, J. W. 1982. Effect of spur defoliation and ringing during bloom on fruiting, fruit mineral level, and net photosynthesis of 'Golden Delicious' apple. *J. Am. Soc. Hort. Sci.* 107(6):1182–1186.

Forshey, C. G. and Elfving, D. C. 1989. The relationship between vegetative growth and fruiting in apple trees. *Hortic. Rev.* 11:229-287.

Gardiner, J. 2000. Using reflective mulch for light management in orchards. Available at: http://www.omafra.gov.on.ca/english/crops/facts/orchard_mulch.htm (accessed June 6, 2022).

Gildemacher, P., Heijne, B., Silvestri, M., Houbraken, J., Hoekstra, E., Theelen, B. and Boekhout, T. 2006. Interactions between yeasts, fungicides and apple fruit russeting. *FEMS Yeast Res.* 6(8):1149–1156.

Goffinet, M. C. and Lakso, A. N. 2017. Positional effects on retention potential within the spur of apple (Malus× domestica) with and without competition with other fruit on the spur. *Acta Hortic.* 1177:169–176.

Goffinet, M. C., Robinson, T. L. and Lakso, A. N. 1995. A comparison of 'Empire' apple fruit size and anatomy in unthinned and hand-thinned trees. *J. Hortic. Sci.* 70(3):375–387.

Greene, D. W. 1993. Effects of GA_4 and GA_7 on flower bud formation and russet development on apple. *J. Hortic. Sci.* 68(2):171–176.

Greene, D. W. 2002. Preharvest drop control and maturity of 'Delicious' apples as affected by aminoethoxyvinylglycine (AVG). *J. Tree Fruit Prod.* 3(1):1–10.

Greenhalgh, W. J. and Godley, G. L. 1976. Studies of fruit shape in apples: A survey of factors influencing the development of 'typiness' of the cultivar Red Delicious. *Aust. J. Exp. Agric.* 16(81):592–595.

Hampson, C. R., Sanford, K. and Cline, J. 2002. Preferences of Canadian consumers for apple fruit size. *Can. J. Plant Sci.* 82(1):165–167.

Harker, F. R., Kupferman, E. M., Marin, A. B., Gunson, F. A. and Triggs, C. M. 2008. Eating quality standards for apples based on consumer preferences. *Postharvest Biol. Technol.* 50(1):70–78.

He, L. and Schupp, J. 2018. Sensing and automation in pruning of apple trees: A review. *Agronomy* 8(10):211. https://doi.org/10.3390/agronomy8100211.

Heinicke, D. R. 1964. The micro-climate of fruit trees: III. The effect of tree size on light penetration and leaf area in Red Delicious apple trees. *Proc. Am. Soc. Hort. Sci.* 85.

Hyson, D. A. 2011. A comprehensive review of apples and apple components and their relationship to human health. *Adv. Nutr.* 2(5):408–420.

Iglesias, I., Salvia, J., Torguet, L. and Cabús, C. 2002. Orchard cooling with over tree microsprinkler irrigation to improve fruit colour and quality of 'Topred Delicious' apples. *Sci. Hort.* 93(1):39–51.

Jacob-Wilk, D., Holland, D., Goldschmidt, E. E., Riov, J. and Eyal, Y. 1999. Chlorophyll breakdown by chlorophyllase: Isolation of the chlase gene from ethylene-treated Citrus fruit and its regulation during development. *Plant J.* 20(6): 653–661.

Jaeger, S. R., Andani, Z., Wakeling, I. N. and MacFie, H. J. H. 1998. Consumer preferences for fresh and aged apples: A cross-cultural comparison. *Food Qual. Prefer.* 9(5):355–366.

Jaeger, S. R., Antunez, L., Ares, G., Johnston, J. W., Hall, M. and Harker, F. R. 2016. Consumers' visual attention to fruit defects and disorders: A case study with apple images. *Posharvest Biol. Technol.* 116:36–44.

Khanal, B. P., Imoro, Y., Chen, Y. H., Straube, J. and Knoche, M. 2021. Surface moisture increases microcracking and water vapour permeance of apple fruit skin. *Plant Biol. (Stuttg)* 23(1):74–82.

King, M. C. and Cliff, M. A. 2002. Development of a model for prediction of consumer liking from visual attributes of new and established apple cultivars. *J. Am. Pomol. Soc.* 4:23–229.

Kon, T. M. and Schupp, J. R. 2018. Apple crop load management with special focus on early thinning strategies: A USA perspective. *Hort. Rev.* 46:255–298.

Lakso, A. N. 1994. Apple: Handbook of environmental physiology of fruit crops. In: Schaffer, B. and Andersen, P. C. (Eds.), *Handbook of Environmental Physiology of Fruit Crop*. London: CRC Press, pp. 3–42.

Lakso, A. N. 2003. Water relation in apples. In: Ferree, D. C. and Warrington, I. J. (Eds.), *Apples: Botany, Production and Uses*. Wallingford, UK: CABI Press, pp. 168–181.

Larson, J. E., Kon, T. M. and Malladi, A. 2022. Apple fruitlet abscission mechanisms. *Hort. Rev.* 49:243–274.

Lin-Wang, K. U. I., Micheletti, D., Palmer, J., Volz, R., Lozano, L., Espley, R., Hellens, R. P., Chagne, D., Rowan, D. D., Troggio, M., Iglesias, I. and Allan, A. C. 2011. High temperature reduces apple fruit colour via modulation of the anthocyanin regulatory complex. *Plant Cell Environ.* 34(7):1176–1190.

Lindow, S. E., Desurmont, C., Elkins, R., McGourty, G., Clark, E. and Brandl, M. T. 1998. Occurrence of indole-3-acetic acid-producing bacteria on pear trees and their association with fruit russet. *Phytopathology* 88(11):1149–1157.

Lordan, J., Wallis, A., Francescatto, P. and Robinson, T. L. 2018. Long-term effects of training systems and rootstocks on 'McIntosh' and 'Honeycrisp' performance, a 15-year study in a northern cold climate – Part 1: Agronomic analysis. *Hort.Sci.* 53(7):968–977.

Macnee, N. C., Rebstock, R., Hallett, I. C., Schaffer, R. J. and Bulley, S. M. 2020. A review of current knowledge about the formation of native peridermal exocarp in fruit. *Funct. Plant Biol.* 47(12):1019–1031.

Malladi, A. 2020. Molecular physiology of fruit growth in apple. *Hort. Rev.* 47:1–42.

Marcelle, R. 1995. Mineral nutrition and fruit quality. *Acta Hortic.* 383:219–226.

Marini, R. P., Barden, J. A., Cline, J. A., Perry, R. L. and Robinson, T. 2002. Effect of apple Rootstocks on average Gala fruit weight at four locations after adjusting for crop load. *J. Am. Soc. Hortic. Sci.* 127(5):749–753.

Marini, R. P. and Fazio, G. 2018. Apple rootstocks: History, physiology, management, and breeding. *Hort. Rev.* 45:197–312.

Matile, P., Hörtensteiner, S. and Thomas, H. 1999. Chlorophyll degradation. *Annu. Rev. Plant Physiol. Plant Mol. Biol.* 50(1):67–95.

McArtney, S., Obermiller, J. D. and Green, A. 2007. Prohexadione-Ca reduces russet and does not negate the efficacy of GA4+7 sprays for russet control on 'Golden Delicious' apples. *Hortscience* 42(3):550–554.

McClure, K. A. and Cline, J. A. 2015. Mechanical blossom thinning of apples and influence on yield, fruit quality and spur leaf area. *Can. J. Plant Sci.* 95(5):887–896.

Meyers, T. and Weidenhamer, M. 1966. *Homemakers' Use of and Opinions About Selected Fruits and Fruit Products* (Vol. 765). United States Department of Agriculture, Statistical Reporting Service.

Miller, S. S. 1988. Plant bioregulators in apple and pear culture. *Hort. Rev.* 10:309–401.

Mol, J. N., Jenkins, G. I., Schafer, E., Weiss, D. J. and Walbot, V. 1996. Signal perception, transduction, and gene expression involved in anthocyanin biosynthesis. *Crit. Rev. Plant Sci.* 15(5–6):525–557.

Mupambi, G., Valverdi, N. A., Camargo-Alvarez, H., Reid, M., Kalcsits, L., Schmidt, T., Castillo, F. and Toye, J. 2021. Reflective groundcover improves fruit skin color in 'Honeycrisp' apples grown under protective netting. *hortTechnology* 31(5):607–614.

Musacchi, S. and Greene, D. 2017. Innovations in apple tree cultivation to manage crop load and ripening. In: *Achieving Sustainable Cultivation of Apples*. Cambridge, UK: Burleigh Dodds Science Publishing, pp. 195–238.

Musacchi, S., Knerl, A., Sheick, R. and Serra, S. 2017, September. Cosmic crisp TM: Optimization of rootstock, training system, and pruning technique. In: *ASHS Annual Conference ASHS* (Vol. 2017).

Musacchi, S. and Serra, S. 2018. Apple fruit quality: overview on pre-harvest factors. *Sci. Hort.* 234:409–430.

Naor, A. 2006. Irrigation scheduling and evaluation of tree water status in deciduous orchards. *Hort. Rev.* 32:111–165.

Nava, G., Dechen, A. R. and Nachtigall, G. R. 2007. Nitrogen and potassium fertilization affect apple fruit quality in Southern Brazil. *Commun. Soil Sci. Plant Anal.* 39(1–2):96–107.

Neilsen, G. H., Hogue, E. J. and Meheriuk, M. 1999. Nitrogen fertilization and orchard floor vegetation management affect growth, nutrition and fruit quality of 'Gala' apple. *Can. J. Plant Sci.* 397–385.

Neilsen, G. H. and Neilsen, D. 2003. Nutritional requirements of apple. In: Ferree, D. C. and Warrington, I. J. (Eds.), *Apples, Botany, Production and Uses*. Oxon, UK: CABI Publishing, pp. 267–302.

Neri, D. and Sansavini, S. 2019. Planting and training systems, pruning and fruiting control. In: *Principles of Modern Fruit Science*. Belgium: Illinois State Historical Society, p. 421.

Nicolai, B. M., Beullens, K., Bobelyn, E., Peirs, A., Saeys, W., Theron, K. I. and Lammertyn, J. 2007. Nondestructive measurement of fruit and vegetable quality by means of NIR spectroscopy: a review. *Postharvest Biol. Technol.* 46(2):99–118.

Nock, J. F., Watkins, C. B., James, H., Reed, N. and Oakes, R. L. 2009. Preharvest application of 1-methylcyclopropene (1-MCP) to control fruit drop of apples, and its effects on postharvest quality. *Act. Hort.* 877:365–374.

Ontario Ministry of Agriculture, Food and Rural Affairs. 2021a. Crop protection guide for apples: Publication 360A. Available at: http://www.omafra.gov.on.ca/english/crops/pub360/pub360A.pdf (accessed April 14, 2022).

Ontario Ministry of Agriculture, Food and Rural Affairs. 2021b. Soil management, fertilizer use, crop nutrition and cover crops for fruit production. Available at: http://www.omafra.gov.on.ca/english/crops/hort/soil_fruit.htm#Magnesium (accessed June 10, 2022).

Palmer, J. W. 1992. Effects of varying crop load on photosynthesis, dry matter production and partitioning of Crispin/M.27 apple trees. *Tree Physiol.* 11(1):19–33.

Palmer, J. W., Wunsche, J. N., Meland, M. and Hann, A. 2002. Annual dry-matter production by three apple cultivars at four within-row spacings in New Zealand. *J. Hortic. Sci. Biotechnol.* 77(6):712–717.

Penn State University Extension. 2022. *2022-2023 Penn State Tree Fruit Production Guide*. University Park, PA, p. 542.

Robinson, T. L. 2007. Recent advances and future directions on orchard planting systems. *Acta Hortic.* 732:367–381. https://doi.org/10.17660/ActaHortic.2007.732.57.

Sansavini, S., Costa, G., Gucci, R., Inglese, P., Ramina, A., Xiloyannis, C. and Desjardins, Y. (Eds.). 2019. *Principles of Modern Fruit Science*. Leuven, Belgium: Illinois State Historical Society, p. 421.

Scalisi, A., McClymont, L., Underwood, J., Morton, P., Scheding, S. and Goodwin, I. 2021. Reliability of a commercial platform for estimating flower cluster and fruit number, yield, tree geometry and light interception in apple trees under different rootstocks and row orientations. *Comput. Electron. Agric.* 191:106519.

Sharples, R. O. 1980. The influence of orchard nutrition on the storage quality of apples and pears grown in the United Kingdom. In: Atkinson, D., Jackson, J. E., Sharples, R. O. and Waller, W. M. (Eds.), *Mineral Nutrition of Fruit Trees*. London, England: Butterworths, pp. 17–28.

Solomakhin, A. A. and Blanke, M. M. 2010. Mechanical flower thinning improves the fruit quality of apples. *J. Sci. Food Agric.* 90(5):735–741.

Stiles, W. C. 1994. Phosphorus, potassium, magnesium, and sulfur soil management. In: *Tree Fruit Nutrition*. Yakima, WA: Good Fruit Grower.

Tagliavini, M. 2019. Mineral nutrition. In: Sansavini, S., Costa, G., Gucci, R., Inglese, P., Ramina, A., Xiloyannis, C. and Desjardins, Y. (Eds.), *Principles of Modern Fruit Science*. Leuven, Belgium: Illinois State Historical Society, p. 421.

Takos, A. M., Jaffe, F. W., Jacob, S. R., Bogs, J., Robinson, S. P. and Walker, A. R. 2005. Light induced expression of a MYB gene regulates anthocyanin biosynthesis in red apples. *Plant Physiol.* 142(3):1216–1232.

Taylor, B. K. 1975. Reduction of apple skin russeting by gibberellin A4+7. *J. Hortic. Sci.* 50(2):169–172.

Taylor, D. R. and Knight, J. N. 1986. Russeting and cracking of apple fruit and their control with plant growth regulators. *Acta Hort.* 179:819–820.

Tromp, J. 2000. Flower-bud formation in pome fruits as affected by fruit thinning. *Plant Growth Regul.* 31(1/2):27–34.

Tromp, J., Webster, A. D. and Wertheim, S. J. (Eds.). 2005. *Fundamentals of Temperate Zone Fruit Production*. Leiden, The Netherlands: Backhuys Publishers.

Tworkoski, T., Fazio, G. and Glenn, D. M. 2016. Apple rootstock resistance to drought. *Sci. Hort.* 204:70–78.

Valverdi, N. A. and Kalcsits, L. 2021a. Apple rootstock genotype affects scion responses to water limitations under field conditions. *Acta Physiol. Plant.* 43(7):1–14.

Valverdi, N. A. and Kalcsits, L. 2021b. Rootstock affects scion nutrition and fruit quality during establishment and early production of 'Honeycrisp' apple. *Hortscience* 56(2):261-269.

Wagenmakers, P. S. and Callesen, O. 1995. Light distribution in apple orchard systems in relation to production and fruit quality. *J. Hortic. Sci.* 70(6):935–948.

Wang, N., Jiang, S., Zhang, Z., Fang, H., Xu, H., Wang, Y. and Chen, X. 2018. *Malus sieversii*: The origin, flavonoid synthesis mechanism, and breeding of red-skinned and red-fleshed apples. *Hortic. Res.* 5:70.

Wang, Y., Dai, M., Cai, D., Zhang, S. J. and Shi, Z. 2016. A review for the molecular research of russet/semi-russet of sand pear exocarp and their genetic characters. *Sci. Hortic.* 210:138–142.

Wang, Y., Li, W., Xu, X., Qiu, C., Wu, T., Wei, Q., Ma, F. and Han, Z. 2019. Progress of apple rootstock breeding and its use. *Hortic. Plant J.* 5(5):183–191.

Warrington, I. J., Fulton, T. A., Halligan, E. A. and De Silva, H. N. 1999. Apple fruit growth and maturity are affected by early season temperatures. *J. Am. Soc. Hortic. Sci.* 124(5):468–477.

Watkins, C. B. 2006. The use of 1-methylcyclopropene (1-MCP) on fruits and vegetables. *Biotechnol. Adv.* 24(4):389–409.

Watkins, C. B. 2008. Overview of 1-methylcyclopropene trials and uses for edible horticultural crops. *Hortscience* 43(1):86–94.

Watkins, C. B. 2017. Advances in postharvest handling and storage of apples. In: *Achieving Sustainable Cultivation of Apples*. Cambridge, UK: Burleigh Dodds Science Publishing, pp. 337–368.

Watkins, C. B., Nock, J. F. and Whitaker, B. D. 2000. Responses of early, mid and late season apple cultivars to postharvest application of 1-methylcyclopropene (1-MCP) under air and controlled atmosphere conditions. *Postharvest Biol. Technol.* 19(1):17–32.

Webster, A. D. and Wertheim, S. J. 2003. Apple rootstocks. In: Ferree, D. C. and Warrington, I. J. (Eds.), *Apples: Botany, Production and Uses*. Wallingford, UK: CABI Press, pp. 91–124.

Weeks, W. D., Southwick, F. W., Drake, M. and Olanyk, G. W. 1965. Relation differential N and K fertilization to tree performance, fruit quality and storage disorders of delicious apples. *Mass. Exp. Stn. Bull.* 552.

Westwood, M. N., Batjer, L. P. and Billings, H. D. 1967. Cell size cell number and fruit density of apples as related to fruit size position in cluster and thinning method. *Proc. Am. Soc. Hort. Sci.* 91:51–62.

Westwood, M. N. and Burkhart, D. J. 1968. Climate influences shape of Delicious. *Am. Fruit Grower* 88(6):26.

Williams, M. W. 1979. Chemical thinning of apples. *Hort. Rev.* 1:270–300.

Williams, M. W. and Edgerton, L. J. 1981. *Fruit thinning of apples and pears with chemicals* (No. 1474-2021-427).

Winkler, A., Athoo, T. and Knoche, M. 2022. Russeting of fruits: Etiology and management. *Horticulturae* 8(3):231. https://doi.org/10.3390/horticulturae8030231.

Wismer, P. T., Proctor, J. T. A. and Elfving, D. C. 1995. Benzyladenine affects cell division and cell size during apple fruit thinning. *J. Am. Soc. Hortic. Sci.* 120(5):802–807.

Wünsche, J. N. and Ferguson, I. B. 2010. Crop load interactions in apple. *Hort. Rev.* 31:231–290.

Yuan, R. and Carbaugh, D. H. 2007. Effects of NAA, AVG, and 1-MCP on ethylene biosynthesis, preharvest fruit drop, fruit maturity, and quality of 'Golden Supreme' and 'Golden Delicious' apples. *Hortscience* 42(1):101–105.

Yuan, R. and Li, J. 2008. Effect of sprayable 1-MCP, AVG, and NAA on ethylene biosynthesis, preharvest fruit drop, fruit maturity, and quality of 'Delicious' apples. *Hortscience* 43(5):1454–1460.

Zhang, J., Serra, S., Leisso, R. S. and Musacchi, S. 2016. Effect of light microclimate on the quality of 'd'anjou' pears in mature open-centre tree architecture. *Biosyst. Eng.* 141:1–11.

Zhang, Q., Karkee, M. and Khot, L. R. 2017. Mechanization and automation for apple production. In: *Achieving Sustainable Cultivation of Apples*. Cambridge, UK: Burleigh Dodds Science Publishing, pp. 279–306.

Chapter 8

Postharvest management of apple quality

Zora Singh, Edith Cowan University, Australia; Vijay Yadav Tokala, The Postharvest Education Foundation, USA; and Mahmood Ul Hasan and Andrew Woodward, Edith Cowan University, Australia

1 Introduction

There is a need to manage practices including harvesting, produce handling, packaging, transport, storage and marketing of the fruits to ensure the supply of acceptable products to the end consumer. The apple fruit is relished for its crispy texture and sweet-tart taste, and it is also rich in dietary fibres, minerals, sugars (including fructose, glucose and sucrose), organic acids (malic acid), antioxidant phytochemicals (quercetin, catechin, phloridzin), vitamins and phenols. The market price of the fruit is decided by the eating quality and attractiveness of the fruit such as size, shape, colour and free from blemishes and internal disorders. The fruit ripening process undergoes a series of irreversible biochemical and physiological reactions, which cause respiration and ethylene production changes, cell wall degradation, loss of chlorophyll, moisture loss and several other reactions, eventually leading to postharvest senescence (Taiz et al., 2015; Musacchi and Serra, 2018). Depending on the respiration rates during the ripening process and their patterns, the fruits are of two types: climacteric and non-climacteric. Climacteric fruits show a distinct rise in respiration rates during the ripening, which are referred to as climacteric

http://dx.doi.org/10.19103/AS.2023.0127.11

peaks. The rise in ethylene biosynthesis in the fruit occurs simultaneously or just after the rise in respiration rates (Tokala et al., 2021a,b). Based on the respiration pattern, the apple fruit is classified as a climacteric fruit that is extremely sensitive to exposure to ethylene and in turn exhibits large amounts of ethylene during ripening (Tokala et al., 2021a). Ethylene plays an important role in upregulating the ripening-associated changes, ultimately leading to quality deterioration and postharvest senescence (Iqbal et al., 2017). Apple fruit experience high rates of moisture loss during storage depending upon its genotype, which results in loss of firmness, increased weight loss and skin shrivelling, thereby declining their marketability (Fig. 1). The postharvest anabolic and catabolic reactions in the fruit result in significant changes in the texture, colour, flavour, aroma and nutritional profile.

The postharvest management of the apple fruit focuses on maintaining the quality during storage and the supply chain. To downregulate the rates of ripening-associated physiological and biochemical changes causing undesirable effects on the fruit quality and controlling the development of diseases and physiological disorders during the postharvest period, the postharvest management of apple fruit must be carried out with a prime focus on the several aspects including harvesting at the precise maturity and proper handling to minimize impact bruises and mechanical injuries.

Depending upon the different quality standards determined by the market (domestic or international) and the possible use of the fruit after harvest, maturity indices have been designated. The right stage of maturity of the apple fruit is generally determined by different factors for easy and rapid assessment such as days from full bloom, soluble solids, titratable acidity (TA), sugar-acid ratio (ripening index), fruit firmness, starch content, peel and flesh colour (Kingston, 1992; Watkins, 2003; Abi Tarabay et al., 2018). The maturity stage of

Figure 1 Bravo™ apple fruit shows clear signs of moisture loss, shrivelling and deterioration of cosmetic appearance and fruit quality following a long storage period.

the fruit at harvest has a significant effect on the storage quality of the fruit. The possible duration of storage with optimum fruit quality can assist in extending storage life if the fruit is harvested at the optimum physiological maturity stage (Shafiq et al., 2011a,b). However, if the fruit is harvested at a too early stage, the colour and flavour of the cultivar fail to develop and become susceptible to physiological disorders such as bitter pit and superficial scald. On the other hand, fruit harvested at the fully mature stage was observed to be less firm and more susceptible to mechanical damage, physiological disorders and diseases (Watkins, 2003). The impact of bruises and mechanical injuries are mainly due to improper handling and can occur during any stage of the supply chain from harvest to consumer. Funt et al. (1999) reported that nearly 40% of bruising occurs due to rough handling in the fields, 40% during grading and the other 20% during transport. The bruise susceptibility of the fruit in general increases at later stages of fruit maturity and thereby also increases the chances for pathological infections. Training the fruit pickers with the right harvest methods, using appropriate harvesting instruments such as picking bags, secateurs and bins, proper packing and transport to reduce the impact on fruit and gentle handling can significantly reduce these losses (Abi Tarabay et al., 2018).

Various techniques have been tested to extend the storage life of the apple fruit. The most common way is to use cold storage to downregulate the activities of various enzymes and thereby retard the undesirable physiological and biochemical changes leading to senescence (Gross et al., 2016). Most of the ripening changes (directly and indirectly) are associated with the ethylene action in fruit. Ethylene induces several oxidative processes and the formation of different reactive oxygen species (ROS), which are associated with the degradation of a range of bioactive compounds in the cell (Masia, 1998a; Steinite et al., 2004). On exposure to exogenous ethylene treatment, the climacteric fruits exhibit intensified internal ethylene production in an autocatalytic way and accelerate the ripening and senescence process (Barry et al., 2000). The apple fruit is generally categorized as high ethylene (>100 $\mu L\ kg^{-1}\ h^{-1}$) producing fruit and is observed to be highly sensitive to ethylene exposure (Burg and Burg, 1962; Abeles et al., 2012). Proper storage conditions and different postharvest treatments to inhibit ethylene action are essential mechanisms to slow down the ripening-associated undesirable changes and extend the storage life while maintaining optimum fruit quality (Gross et al., 2016; Tokala et al., 2021a). The negative influence of ethylene can be inhibited by blocking ethylene biosynthesis and antagonizing the ethylene action. 1-Aminoethoxyvinylglycine (AVG), amino oxyacetic acid (AOA) and methoxyvinylglycine (MVG) have been regarded as inhibitors of ACC synthase (ACS) activity, while α-amino isobutyric acid (AIB) and ethanol suppressed *ACC oxidase* (ACO) activity (Martínez-Romero et al., 2007). Abeles et al. (2012) reported that a higher concentration of carbon dioxide (CO_2) in the storage reduced autocatalytic ethylene production in several climacteric fruits including apples. The antagonizing of the ethylene action is possible by irreversible blocking of the ethylene receptors at the cellular level. The details about some of the ethylene antagonist compounds and their role in blocking ethylene action have been reviewed by Sisler et al. (2006), who noted that 1-methylcyclopropene (1-MCP) was comparatively the most effective cyclopropene (CP) in antagonizing the ethylene action in fresh horticultural produce. Management of the storage environment is a critical factor in the extension of storage life and in delaying the deterioration of the fruit quality (Domínguez et al., 2016). Cold storage, controlled atmosphere (CA) and modified atmosphere packing (MAP) are commonly employed methods for the long-term storage of fruits. The inclusion of facilities such as photocatalytic oxidization or ozone in the storage environment had a synergistic effect in prolonging the storage life of apple fruit (Tokala et al., 2021b). In the present chapter, the different postharvest practices along the supply chain of apple fruit will be discussed with a focus on recent interventions.

2 Pre-storage

2.1 Pre-cooling

Temperature management in fruit during postharvest period is one of the most principal factors in extending storage life by modulating physiological and biochemical processes (Mercier et al., 2017). The fruit at harvest usually carries field heat which accelerates metabolic processes and also the growth of microorganisms (Ambaw et al., 2017; Duan et al., 2020). The continuing physiological activities promote moisture loss from the fruit, which further causes a loss of weight and adversely impacts quality, including the texture (Duan et al., 2020; Lufu et al., 2019). The intensity of moisture loss depends on the genotype and vapour pressure deficit, which is the difference between saturation vapour pressure and actual air vapour pressure, and it is strongly affected by temperature and relative humidity in the storage environment (Fanourakis et al., 2013; Wills and Golding, 2016). Pre-cooling is attempted just after fruit harvest before transporting into cold storage or market to remove field heat from the fruit and cool the produce by up to 87% of the initial temperature. Room cooling, forced air cooling, hydro cooling, liquid ice-cooling and vacuum cooling are the common pre-cooling methods used in fresh horticultural produce (Gross et al., 2016). Room cooling, forced air cooling and hydro cooling methods are often used in the apple industry, and each method has its pros and cons (Table 1).

2.1.1 Room cooling

In this relatively slow way of pre-cooling, the produce is put into a cold room and cold air is circulated around the cartons, crates or bins of apples. For quick

Table 1 Comparison of different pre-cooling methods for apples

	Room cooling	Forced air cooling	Hydro cooling
Cooling time range (h)	20 to 100	1 to 10	<1
Water contact	No	No	Yes
Energy efficiency	Low	Low	High
Advantages	Easy, low maintenance	High efficiency, faster than room cooling	Fast and uniform cooling, no moisture loss from fruits, less expensive, include chemical treatments to control storage diseases
Disadvantages	Uneven cooling, slow process	Uneven cooling	High chances of microbial cross-contamination, water being circulated must be disinfected

cooling of produce, it is essential to keep passable space between the stacks inside the refrigerated rooms. Produce in vented boxes is cooled faster than in un-vented boxes. About four to six inches (10–15 cm) of space between the stacks is sufficient to let the cold air circulate around the individual boxes or pallet loads. The stacks should be about one pallet width in-depth, and fans should be fitted to flow the cool air in the entire room. It is essential to monitor the temperature of produce within the packages at various locations in the room to check the status of produce being cooled at the desired temperature (Thompson et al., 2002).

2.1.2 Forced air cooling

This method is an improvement of traditional room cooling where the cooling efficiency is increased by installing fans to enhance the airflow rates through the produce (Ambaw et al., 2017). The cool air is forced through the vents of the packed and palletized produce. The forced air cooling reduces the cooling time of a batch from days to hours. The air velocity and temperature are significant factors for efficient cooling and maintaining fruit quality. The higher rates of airflow and very low air temperatures may enhance moisture loss from fruits and result in disorders such as wilting, shrivelling and chilling injury (Delele et al., 2008).

2.1.3 Hydrocooling

In this method, the produce is cooled by direct contact with cold water and is among the most inexpensive cooling techniques. Due to its higher relative density, water is a better heat transfer medium as compared to air. It provides uniform and faster cooling with no moisture loss from the fruit during the process (Mitchell, 1992). The water being used for hydro cooling should be from a clean source as it can otherwise spread disease inoculum and human pathogens to fresh fruit. The water can be chlorinated to reduce the level of decay-causing microorganisms (Thompson et al., 2008). The immersion hydro coolers are large, shallow, rectangular tanks with submerged conveyors and moving water. The cool and sanitized water is constantly circulated into the tanks. The warm produce in the crates/boxes is loaded on one side of the conveyor belt and submerged until the desired temperature is reached. The duration for which the produce is kept submerged depends on the initial produce temperature and the diameter of the horticultural produce (Thompson et al., 2002). A shower-type hydro cooler distributes cold water on the produce from the top of the container. The water collected at the bottom is cooled and sanitized to recirculate back to the top of the cooler.

Pre-cooled produce tends to have lower moisture loss during the postharvest period. This directly affects the financial returns as the fruit in the market are typically sold based on its weight, and it reduces skin shrivelling and wilting during the storage period. The fruit's firmness and crispness are imperative sensory quality attributes in apple fruit. The activity of fruit softening enzymes including polygalacturonases, pectin methyl esterases and pectin lyases associated with loss of firmness is directly proportional to the core temperature of the fruit (Wills and Golding, 2016). Therefore, the removal of field heat facilitates maintaining the firmness of the apple fruit. Pre-cooling instantly after harvest delayed the fruit softening in 'Royal Gala' and 'Cox's Orange Pippin,' possibly by delaying autocatalytic ethylene production (Johnston et al., 2005). Luo et al. (2018) reported that delayed pre-cooling in apple fruit reduced the quality and storage life caused by changes in levels of 78 proteins which are responsible for cell wall degradation, amino acids, lipids and cell vesicle transport. Contrarily, delayed pre-cooling exhibits a significant reduction in soft scald and chilling injury disorders in 'Honey Crisp' and 'Ambrosia' apple fruit, respectively (DeLong et al., 2004; Luo et al., 2018).

2.2 Edible coatings

Edible coatings exhibit a significant influence on maintaining nutritional and physicochemical properties, modulating gaseous exchange, protecting against microbial decay due to their antimicrobial properties and ensuring antioxidant potential during storage and transportation. Edible coatings emulate the principle of the modified atmosphere by surrounding the fruit surface and manipulating the rates of respiration and ethylene production (Maringgal et al., 2020). The edible coatings are thin layers of edible food-safe materials spread over the fruit surface to create a barrier between the fruit and the external environment (Baldwin et al., 2011). The ideal edible coating would retard respiration, moisture loss and the senescence process (Jongsri et al., 2016). Mahajan et al. (2014) stated that most edible coatings are transparent, without any unpleasant taste and odour, with selective permeability to gases and aroma-volatile compounds. The edible coatings are a harmless and effective way to reduce microbial spoilage when stored at 4°C, while preserving most of the quality and nutritional parameters (Guerreiro et al., 2017). The edible coating is usually developed from plant and animal products, with the main ingredients being polysaccharides, proteins and lipids. It can also be a combination of plasticizers, emulsifiers and active antioxidants. Edible coatings can carry potential additives, and many findings reported the positive effects of edible coatings when blended with essential oils and plant extracts with antimicrobial effects (Maringgal et al., 2020). A range of essential oils including thyme, lemongrass, cinnamon and clove have been proven to have antimicrobial properties as well as being useful in the prevention or reduction of postharvest diseases (Oliveira et al., 2019). Singh and Packirisamy (2022) applied silk fibroin nano-fibre coating to apple fruit and observed that texture, quality and taste were well maintained. The hot-melt paraffin wax was primarily used to coat apple fruit since the 1930s in China. However, the legislation in many countries does not allow paraffin wax to be used as edible coatings (Dhall, 2013). There are several materials used as edible coatings such as *Aloe vera* (Khan et al., 2019), sodium alginate (Kadzińska et al., 2020), chitosan (Khalifa et al., 2017), β-glucan stearic acid ester (Ali et al., 2019), soybean and gum Arabic (El-Anany et al., 2009), tragacanth gum (Jahanshahi et al., 2018) and edible rice starch (Thakur et al., 2019) help in extending the storage time for apple fruit (Table 2). Recent innovations in nanotechnology coatings have improved efficacy in maintaining fruit quality and prolonging the storage life of apples. Fruit treated with 0.5% nano-chitosan coatings regulated the climacteric ethylene production and delayed the ripening of the 'Golab Kohnaz' apple. Coated apples maintained the fruit firmness, colour, total soluble solids and TA throughout the storage, compared to control fruit (Gardesh et al., 2016). Jahanshahi et al. (2018) reported that the tragacanth (*Astragalus gummifer*) gum coating extended shelf life, maintained fruit quality and reduced the microbial load in 'Red Delicious' and 'Golden Delicious' fruit. Similarly, Thakur et al. (2019) observed delayed ripening in the 'Cripps Pink' fruit coated with starch-based edible coatings and stored at low temperatures. While current and emerging research into the application of nanoparticle coatings is very promising, there are challenges from a regulatory perspective. There are many instances of countries allowing various types of nanoparticles, but food safety must be determined before their application (Arts et al., 2015).

Table 2 Effect of postharvest edible coatings alone or in combination with other coating material on postharvest quality and storability of apple fruit

Edible coating	Cultivar	Coating concentration	Concentration of other coatings	Storage conditions	Inference	References
Aloe vera	-	10%	2% calcium chloride	75 days in cold storage	Reduced weight loss and decay; retained ascorbic acid and SSC; maintained firmness; conserved taste quality	Khan et al. (2019)
	'Hongro'	50%	0.5% cystine	16 days at 4°C	Inhibited browning; reduced softening; delayed weight loss; decreased microbial growth; maintained quality	Song et al. (2013)
	'Fuji'	50%	1% lemon oil and 0.1% HPMC	9 days at 4 ± 1°C with 90 ± 5% RH	Retained SSC, TA; delayed browning; restricted fruit softening	Farina et al. (2020)
Sodium alginate	-	1.5%	0.2% rapeseed, coconut and hazelnut oil	-	Retained visual quality; maintained fruit colour	Kadzińska et al. (2020)
Chitosan	'Anna'	2%	1% olive leaves and 2% olive pomace extracts	35 days at 4 ± 1°C	Improved nutritional quality; reduced weight loss and fruit decay; delayed changes in anthocyanins, total phenolics, flavonoids, carotenoids, and chlorophyll content	Khalifa et al. (2017)
Wheat straw arabinoxylan (AX) and oat bran β-glucan (BG)	'Royal Delicious'	1-2%	-	60 days at 22 ± 2°C; 65-85% RH	Lower cytotoxicity; reduced weight loss, loss of aroma and decay; higher firmness and delayed ripening index and polyphenol oxidase activity	Ali et al. (2019)

Gum Arabic	'Anna'	15%	-	60 days at 0°C with 90–95% RH	Reduced weight loss; maintained higher firmness and visual quality; delayed SSC and acidity	El-Anany et al. (2009)
Rice starch	'Cripps Pink'	2.5%	2% ι-carrageenan	5 weeks at 20°C with 55 ± 2% RH and 10 weeks at 5°C with 90 ± 2% RH + 1 day at 20°C	Maintained higher firmness; reduced mass loss; non-significant effect on SSC and acidity; delayed colour change; improved visual quality; no effect on nutritional quality	Thakur et al. (2019)

HPMC = hydroxypropyl methylcellulose; RH = relative humidity; SSC = soluble solid contents; SSC:TA = sugar:acid ratio.

3 Storage

3.1 Plant growth regulators

Plant growth regulators promote, retard and inhibit several physiological functions in plants. Plant growth regulators can trigger and coordinate a physiological action in combination either synergistically or antagonistically. The effects of plant growth regulator application may often vary among cultivars, locations and applied concentrations (Paulson et al., 2005; Tokala et al., 2020a). Different plant growth regulators have been used to prolong storage life and minimize loss of apple fruit quality in the postharvest phase (Masia et al., 1998b).

The application of daminozide improved peel colour and fruit firmness and downgraded the incidence of watercore disorder in storage in 'Rome' apple fruit (Gianfagna et al., 1989). Previously, Blanpied et al. (1967) reported that daminozide sprays promoted internal browning and shortened the storage life of 'Mcintosh' apples. In addition, the application of α-tocopherol and shikimic acid has enhanced the accumulation of anthocyanins and enhanced storage life in 'Elstar', 'McIntosh' and 'Jonagold' apples (Schmitz and Noga, 1996). Preharvest application of prohexadione-calcium in 'Braeburn' apples reduced ethylene production, delayed starch degradation and loss of fruit firmness as well as TA compared to control fruit in storage (Amarante et al., 2021). Postharvest application of polyamines (spermine and spermidine) through pressure infiltration maintained higher fruit firmness and inhibited the development of chilling injury in 'Golden Delicious' and 'McIntosh' apples (Kramer et al., 1991). Postharvest treatment of melatonin (1 mM L^{-1}) delayed ripening by downregulating the expression of genes implicated in ethylene biosynthesis, reduced weight loss and promoted antioxidant activity in 'Fuji' apples (Onik et al., 2021). Exogenous application of methyl jasmonate improved peel colour, antioxidant capacity, reduced fruit softening and incidence of bitter pit disorder in 'Fuji' and 'Pink Lady' apples (Rudell et al., 2005; Shafiq et al., 2011c, 2013). Nevertheless, the methyl jasmonate application enhanced ethylene synthesis and enhanced colour development in 'Golden Delicious and 'Pink Lady' apples (Fan et al., 1998; Shafiq et al., 2011c, 2013). Hadian-Deljou et al. (2017) have reported that the application of salicylic acid positively impacts the cold storage life and quality of the 'Red Delicious' apple fruit. The 'Cripps Pink' apple fruit exposed to nitric oxide before storage lowered the respiration rates, delayed colour change and lowered flesh browning but did not affect flesh firmness (Steffens et al., 2021).

3.2 Ethylene management

Ethylene is a colourless and odourless gaseous hormone produced by fresh fruit and vegetables during the ripening process (Saltveit, 1999). It has a significant effect on climacteric fruit by initiating and accelerating irreversible physiological processes of ripening, senescence and deterioration. Based on physiological and genetic factors, sensitivity to ethylene gas exposure and amounts of gas emitted by fruit varies from produce to produce (Saltveit, 1999). Ethylene directly or indirectly causes nearly half of all postharvest loss in horticultural crops (Blanke, 2014). Postharvest ethylene management has been suggested as one of the most efficient ways to maintain fruit quality, prolong storage life and reduce losses in fresh horticultural produce (Tokala, 2019).

Endogenous ethylene plays an important role in anthocyanin biosynthesis and the development of the red colour in apple fruit (Whale and Singh, 2007). Some growers use preharvest spray of ethephon to advance the colour development in the fruit by several weeks (Whale et al., 2008). The ethylene treatment provides an opportunity for early ripening when the apples are to be sold in the retail market immediately after the harvest. However, it is a rare practice as it leads to reduced storage life, shorter shelf life and susceptibility to higher fruit loss (Looney, 2018). The long-term storage capacity of apple fruit improves profitability, particularly for long-distance markets, where the ability of the fruit to maintain its fundamental characteristics is essential. Techniques for inhibiting ethylene's negative effects are critical tools for extending storage life and reducing postharvest losses in fresh horticultural produce. The deleterious effects of ethylene can be minimized by two different approaches including inhibition of ethylene biosynthesis and ethylene action inhibition. The key steps in ethylene biosynthesis are the formation of 1-aminocyclopropane-1-carboxylic acid (ACC) from *S*-adenosyl methionine (SAM) and the oxidation of ACC to form ethylene. These steps are catalysed by ACS and ACO enzymes. Martínez-Romero et al. (2007) reported that ethylene biosynthesis is usually blocked with MVG, AVG and AOA, which inhibit ACS activity, by AIB as well as suppression of ACO activity by ethanol. The application of polyamines such as spermine and spermidine reduced the ethylene biosynthesis in the apple fruit by competing with ethylene for SAM (Wang et al., 1993; Valero et al., 2002). The autocatalytic production of ethylene in the apple fruit can be reduced by maintaining higher concentrations of CO_2 in the storage environments (Abeles et al., 2012).

Ethylene antagonists are a group of compounds that bind irreversibly to the receptors in plant cells and block ethylene action. Inhibition of the ethylene action is one of the most effective ways of counteracting the deleterious effects of ethylene during storage in fresh horticultural produce (Sisler et al., 2006). Sisler et al. (2006) investigated the efficacy of different cyclic alkene compounds in inhibiting ethylene action. It was reported that 2,5 norbornadiene (2,5-NBD), *trans* cyclooctene, cyclopentadiene and cyclobutene were efficient in antagonizing the ethylene action but had a pungent, unpleasant smell and required continuous exposure for effective performance. The ethylene antagonistic effect of CP retains for a longer period than other cycloalkene compounds, and a single exposure induces a significant effect for several days. Sisler et al. (2006) reported various possible forms of cyclopropene analogues such as 1-MCP, 3-methylcyclopropene, 3,3-dimethylcyclopropene and 1-ethylcyclopropene inhibited ethylene biosynthesis during the postharvest period. The 1-MCP proved to be more stable in a gaseous state and exhibited its ethylene antagonistic effect for longer durations. 1-MCP was found to be non-toxic and did not leave any detectable residues in fruit or vegetables. After receiving Environmental Protection Agency approval, AgroFresh Inc. developed an encapsulated form of 1-MCP to use on fresh fruit and vegetables with the trade name SmartFresh™ (Watkins, 2006, 2008). Further, the 1-MCP has revolutionized the horticulture industry and its commercial formulations are extensively used by apple growers to prolong storage life.

1-MCP is efficient in downregulating ethylene production, respiration rate, colour change and softening in fresh horticultural produce (Blankenship and Dole, 2003; Watkins, 2006, 2008; Sisler, 2006; Schotsmans et al., 2009; Li et al., 2016; Baswal and Ramezanian, 2021). However, Escribano et al. (2017) reported that the 1-MCP fumigation efficacy varied depending on the maturity stage, cultivars, storage conditions, exogenous ethylene presence, concentrations applied and duration. However, Watkins (2006) reported that 1 μL L^{-1} (18 μM) is the maximum permitted limit as treatment concentration for apple fruit. In addition, the application of the antagonists before the onset of fruit ripening showed better results; however, it varied when applied at different maturity stages (Apelbaum et al., 2008).

In apple and pear fruits, 1-MCP significantly inhibited ethylene production and delayed the steep rise in internal ethylene concentrations associated with climacteric ripening. The duration of the delay or inhibition is determined by the cultivar, storage type and storage time (Watkins, 2008). Several researchers reported that 1-MCP application inhibits/delays colour development in apple peels (Watkins, 2006). Whilst Dauny and Joyce (2002) reported no significant effects of 1-MCP on the apple fruit peel colour, Dauny and Joyce (2002) and Baritelle et al. (2001) claimed that 1-MCP fumigation-maintained cell turgor and reduced the physiological loss of weight.

The 'Redchief Delicious' apple fruit treated with 1-MCP retained higher fruit firmness than control fruit when stored at 20–24°C ambient conditions (Mir et al., 2001). Contrarily, Toivonen and Lu (2005) reported that the 'Sunrise Summer' apple treated with 1-MCP failed to retain fruit firmness which was stored at 15°C. Researchers reported varied responses to the effect of the 1-MCP application on the nutritional parameters of fruits. MacLean et al. (2006) argued that apples treated with 1-MCP exhibited higher levels of antioxidants, whilst Larrigaudière et al. (2004) recorded no significant difference in ascorbic acid and antioxidant levels. Presently, different formulations of 1-MCP are being manufactured by different companies such as Hazel®, AgroFresh and Logfresh® in various delivery forms such as liquid, powder, spray and tablet for commercial application in horticultural crops (AgroFresh, 2022; Logfresh, 2022 and Hazel Technology, 2022). 1-MCP is mostly applied as postharvest treatment after the fruit were harvested, and recently Agrofresh has launched a new formulation Harvista™, which allows preharvest application as a foliar spray in the apple orchard (Watkins, 2015). Singh et al. (2018) discovered new ethylene antagonistic action of 1*H*-cyclopropabenzene (BC) and 1*H*-cyclopropa[*b*] naphthalene (NC) in plants and plant organs. The mechanism of ethylene antagonism of these compounds was reported to be similar and comparable to 1-MCP. BC and NC are more stable compared to 1-MCP in terms of structural differences, with BC being a liquid and NC being a solid at room temperature (Singh et al., 2018). Because of these qualities, BC and NC can be used in a variety of ways, offering attractive alternative options for using ethylene antagonists (Singh et al., 2018). Previously Tokala et al. (2020b) reported that 'Cripps Pink' apple fruit treated with BC and NC fumigation effectively retarded ethylene production and maintained quality during the entire period of storage at low-temperature conditions. Recent studies also revealed similar results for BC and NC fumigation on the quality preservation of apple fruit cultivars. 'Cripps Pink' and 'Granny Smith' in ozonated cold storage and CA storage fitted with photocatalytic oxidation (PCO) setup (Tokala et al., 2021a,b, 2022).

3.3 Storage environments

The augmentation of storage life and mitigation of fruit quality losses during storage is strongly influenced by the conditions of the storage environment including temperature, the composition of oxygen (O_2) and CO_2 and relative humidity (Domínguez et al., 2016; Gross et al., 2016). The apple industry commercially uses cold, CA storage, dynamic CA and MAP for long-term storage and abating postharvest losses in fruit quality. The usage of postharvest application of 1-MCP fumigation in combination with cold, CA, dynamic CA and MAP exhibits synergistic effects on extending storage life and maintaining fruit quality of apple fruit.

3.3.1 Cold storage

Cold storage is one of the traditional and reliable storage methods to prolong the storage life of fresh horticultural produce (Bridgemohan et al., 2021). Fruit are living tissues where respiration and various physiological changes continue even after being separated from the parent plant. Higher temperatures at any point in the supply chain can increase natural degradation processes, resulting in loss of natural colour, flavour, texture, and nutrient levels. Most of the horticultural produce has a Q_{10} temperature coefficient of around two, which means for every 10°C rise in temperature, the rate of enzyme activity causing degradation nearly doubles. Likewise, the enzyme activity is reduced to half when storage temperatures are lowered below 10°C, within the safe storage temperature limits (Bridgemohan et al., 2021). The rate of respiration, ethylene production and water loss are reduced when the produce is stored at low temperatures and high humidity (Kitinoja and Kader, 2002; Keller et al., 2013; Gross et al., 2016). The fruit firmness, nutritional and organoleptic properties of fruit are also maintained during low-temperature storage (Paull, 1999).

The recommended temperatures for storing apple fruit vary among cultivars but, in general, are stored above freezing (0°C) to avoid chilling injury and below 4°C to achieve the maximum benefits of low-temperature storage (Watkins, 2003). In general, the rate of respiration in the apple fruit is greatly reduced during cold storage, but in the case of the apple cultivars susceptible to chilling injury, the respiration rate might increase as a response to the creation of cold stress conditions. Johnson and Ertan (1983) claimed that the 'Idared' apple fruit showed a lower rate of respiration when stored earlier at 0°C in comparison with the fruit stored at 2°C and 4°C; however, respiration rate was observed to increase with progression in storage period. In addition, different physiological disorders including watercore, superficial scald and internal browning commonly occur in apple fruit when the cold storage period is extended for a longer duration (Lurie and Watkins, 2012). Diphenylamine (DPA) and 1-MCP applications have been employed to downregulate the ethylene action and regulate the physiological disorders and extend the storage life of the apple fruit (Isidoro and Almeida, 2006; Jung and Watkins, 2008). These chemicals significantly downregulate ethylene production and α-farnesene, which are responsible for causing superficial scalds on apple fruit (Jung and Watkins, 2008).

3.3.2 Controlled atmospheric storage

The CA method increases the storability of fruits by reducing ethylene production and respiration, thereby delaying other physiological processes linked with the fruit ripening process and postharvest senescence (Keller et al., 2013). At increased CO_2 levels, the autocatalytic biosynthesis of ethylene in climacteric fruits has been reported to be suppressed (Abeles et al., 2012). The gas concentrations are also unfavourable to insect pests and the growth of fungi which helps to reduce pest and disease damage during storage (Kader and Rolle, 2004; Saltveit, 2003). The ideal O_2 and CO_2 concentrations in CA storage vary with fruit crop, species, cultivar, storage temperatures, preharvest circumstances and harvest maturity. Physiological disorders such as internal browning might be aggravated by the stress generated by low O_2

concentrations. Extreme high CO_2 levels cause anaerobic conditions in the fruit resulting in an unpleasant odour and fermented flavour (Thompson, 2010; Gross et al., 2016).

Though CA storage was effective in reducing the physiological disorders during apple storage to an extent, there were different quality issues such as off-flavour development in apples reported during storage. The dynamic controlled atmosphere (DCA) technology has been developed with modifications in the CA storage to address these issues (Wright et al., 2015). The apple can be stored in CA by bringing down the O_2 concentration to the minimum point of lower oxygen limit (LOL), without the occurrence of any quality defects in the fruit even after long-duration storage. The O_2 levels employed in CA during apple storage are still much higher than LOL, which leaves the possibility for quality reduction. However, maintaining O_2 levels to the point of LOL needs constant monitoring throughout the storage period. Various sensor technologies are available around the globe which can constantly monitor the fruit metabolism along the storage period including technologies based on ethanol production (Veltman et al., 2003), chlorophyll fluorescence (Prange et al., 2007) and respiratory quotient (Weber et al., 2015). Mditshwa et al. (2018) exclusively reviewed the mechanisms of different DCA systems, their effects on the apple fruit quality and physiological disorders during long-duration storage. They found that DCA reduced respiration rate and ethylene production and that superficial scald is suppressed as a function of the ethylene reduction. They further concluded though that there is still a risk of high CO_2 and low O_2 injury in DCA-stored fruit, and this negative effect warrants further investigation.

3.4 Other quality management technologies

3.4.1 Ozone

Ozone has been regarded as a powerful oxidizing agent with a great potential property of oxidizing than that hypochlorous acid or chlorine. Manousaridis et al. (2005) stated that ozone can be easily broken into superoxide, hydroxyl, and hydroperoxyl radicals, which can oxidize a wide range of organic and inorganic substances. A wide range of organic compounds such as alcohols, amines, arenes, alkanes and alkenes (including ethylene) are oxidized by ozone to form two carbonyl compounds (Gillies et al., 1988; Tokala et al., 2018). Several studies revealed that exogenous application of ozone either gaseous or aqueous has the potential to reduce the microbial load furthering microbial spoilage on fresh horticultural produce, as well as its effects on physiological and quality parameters during storage (Maletsika et al., 2014; Horvitz and Cantalejo, 2014; Tzortzakis and Chrysargyris, 2017; Tokala et al., 2018, 2021b; Maryam et al., 2021).

To disinfect the fruit surface, aqueous ozone has been reported to be more effective than ozone application in gaseous form. The penetrating capacity and halftime of ozone in the aqueous state, however, are significantly lower than in the gaseous state. In an aqueous state, achieving a higher ozone concentration is generally difficult. A more useful technique to treat large quantities of fruit at once is to use gaseous ozone in sealed rooms (Palou et al., 2001). Earlier studies showed that apple fruit treated with gaseous ozone maintained higher fruit firmness, delayed weight loss, extended storage life and reduced microbial load without any health hazards and phytotoxic signs (Bazarova, 1982; Skog and Chu, 2001; Yaseen et al., 2015). Skog and Chu (2001) also reported that ozone gas application in the low temperature storage extended storage life by oxidizing ethylene in the storage environment but did not have any effect on the endogenous ethylene production by the apple fruit. The combination of ozone with cold (Liew and Prange, 1994) or CA storage conditions (Concha-Meyer et al., 2015) enhanced the bioactive compounds and helped in maintaining fruit quality and extending storage life as compared to independent treatments. Tokala et al. (2021b) have recently reported that apple fruit subjected to different treatments in ozonated cold storage significantly maintained higher sugar levels but resulted in elevated ethylene production during the postharvest period which warrants future investigations (Table 3).

3.4.2 Photocatalytic oxidation

PCO oxidizes a variety of airborne pollutants as well as organic gaseous substances like ethylene to produce water and CO_2. The device works by destroying airborne organic particles as they pass through it using an interaction between ultraviolet-C (UV-C) light and a unique substrate coating (Tokala et al., 2022). Sporadic and inconclusive work has been conducted on the efficacy of PCO in the storage environment to remove ethylene. Tokala (2019) reported that installing PCO in the CA storage exhibited a synergistic effect in downregulating the metabolic activities in 'Cripps Pink' apple fruit than conventional CA storage following 90 days, but the efficiency lessened after 120 days of storage (Table 3).

3.4.3 Cold plasma

Plasma is considered the fourth state of physical matter, and it is created by ionizing a gas with huge quantities of energy. Multiple chemical processes and ultraviolet (UV) radiation occur in a high-energy system including diverse active species such as molecules, atoms, ions, electrons and free radicals (Chen et al., 2020). Muhammad et al. (2018) described that plasma is a unique type of matter in addition to physical interactions such as etching, collision, UV radiation, and

Table 3 Effect of different technologies on postharvest physiology, storage life and quality of apple fruit

Technologies	Cultivar	Concentration	Storage conditions	Inference	References
Ozone	'Royal Gala', 'Golden Delicious', 'Fuji'	0.5 μL L^{-1}	2 months at 21 ± 1°C	Reduced fungal decay; inhibited patulin biosynthesis thereby reducing *Penicillium expansum*	Yaseen et al. (2015)
	'Auksis', 'Gita', 'Ledzenu', 'Elorusskoje Malinovoje'	0.8 ppm	6 months at 0.2 ± 0.1°C, 3.1 ± 0.0% O_2, 0.4 ± 0.0% CO_2 with 90% RH	Maintained fruit firmness; inhibited microbial spoilage; retained sensory quality	Juhnevica-Radenkova et al. (2019)
	'Red Delicious'	60.2–78.7 μg L^{-1}	36 weeks at 4°C	Limited microbial decay; reduced *Listeria innocua* growth; maintained higher firmness	Shen et al (2021)
	'Cripps Pink', 'Granny Smith'	0.1 ± 0.08 μL L^{-1}	90 days at 0 ± 2°C with 90 ± 5% RH	Maintained higher sugar levels; elevated ethylene production	Tokala et al. (2021b)
Photocatalytic oxidation	'Cripps Pink'	Standard as per AiroFresh®	150 days at 3.45 ± 0.45% O_2 and 2.40 ± 0.36% CO_2, 0 ± 1°C with 90 ± 5%	Maintained fruit quality; downregulated ethylene action and respiration rate	Tokala et al. (2022)
UV radiations	'Hongyu'	9.0 kJ m^{-2} UV-C	14 days at 20°C with 85–95% RH	Modulate malate metabolism; improved taste quality; promote early ripening	Onik et al. (2019)

(*Continued*)

Table 3 (*Continued*)

Technologies	Cultivar	Concentration	Storage conditions	Inference	References
	'Red Delicious', 'Golden Delicious'	1.4×10^{-4} w cm^2 for 15 min UV-C	6 months at 1 ± 1°C with 85–95% RH	Lowered pH, soluble solids content and acidity ratio; maintained higher firmness; reduced loss of fruit quality	Hemmaty et al. (2006)

RH = relative humidity; UV = ultraviolet.

the chemical interactions between active species and samples which makes plasma a distinct form of matter. In the case of non-thermal plasma, overall gas temperatures remain low and suitable for the surface sterilization of food produce (Tolouie et al., 2018). Cold plasma with temperatures close to room temperature was effective in inactivating microorganisms on the fruit surface to reduce disease incidence and ensure food security (Ekezie et al., 2017). Chen et al. (2020) reviewed the effects of different plasma treatments on the physical, chemical and physiological quality changes in fresh and fresh-cut produce. Tappi et al. (2019) applied plasma treatment on 'Pink Lady', 'Fuji' and 'Modi' apples which enhanced the firmness and crunchiness of the fruit. Whilst, with increased plasma treatment time, the texture of the 'Red Delicious' apple was negatively affected. Schnabel et al. (2015) stated that the aroma and peel colour of the apple fruit remained unchanged with microwave plasma treatment. Whilst Baier et al. (2015) reported that green apples containing chlorophyll were less susceptible to colour change by plasma than carotenoid-rich peels.

3.4.4 Ultraviolet-C radiation and ultrasound technology

Similar to cold plasma, non-thermal surface sterilization technologies such as UV-C radiation and ultrasound (US) technology have been widely investigated in horticultural produce (Lu et al., 2020; Esua et al., 2020; Maryam et al., 2021). UV-C radiations are electromagnetic waves that have wavelengths between 250 nm and 280 nm and are associated with the disruption of microorganism DNA functioning and integrity, as well as the creation of ROS, ultimately negatively affecting the microbial population on the surface exposed (Darras et al., 2020). Whilst the high-intensity US with low frequencies of 20–100 kHz affect the microbial population by inducing acoustic cavitation in the cells to generate reactive hydroxyl radicals causing microbial inactivation and stimulation of secondary metabolite accumulation in fruits and vegetables (Lu et al., 2020). Though the use of UV-C radiation and US technology have encouraging effects, the possible combination with other technologies for harnessing the beneficial effects on horticultural fresh produce is recommended over individual use (Esua et al., 2020). Apple fruit cv. Hongyu subjected to UV-C treatment at 9.0 kJ m^{-2} markedly modulated malate metabolism, improved taste quality and has shown potential for early promotion of fruit ripening (Onik et al., 2019) (Table 3).

3.5 Postharvest disorders

Apple crop is vulnerable to various disorders, which affects their physiological and organoleptic quality. Postharvest physiological disorders in apple fruit are caused by various factors, from preharvest nutrition management to gas levels of O_2 and CO_2 in storage environments which adversely impact apple

Table 4 Postharvest physiological disorders in apples

Disorder	Symptoms	Susceptible Cultivars	Causes	Control
Bitter pit	Cell structure breakdown during storage, skin hard, sunken and dry brown lesions	'Braeburn', 'Golden Delicious', 'Granny Smith', 'Honey Crisp', 'Goldspur' and 'Cox's Orange Pippin'	Low calcium concentration, loss of xylem functionality, water stress, higher GA concentrations	Preharvest and postharvest calcium treatments, rapid pre-cooling, high humidity and CA storage
Brown core/core flush/core browning	Brown or flesh pink discolouration, tissue softening	'Baldwin', 'Braeburn', 'Bramley's Seedling', 'Cox's Orange Pippin', 'Empire', 'Fuji', 'Honeycrisp', 'McIntosh', 'Granny Smith', 'Red Delicious', 'Rhode Island Greening', 'Topaz'	Early harvesting, delayed pre-cooling, excessive water loss during storage, low calcium content and extended storage duration	Harvest at the right maturity, DPA application and DCA storage
CO_2 injury	Brown epidermal discolouration, moist brown core turns dry and spongy	External symptoms 'Bramleys' Seedling', 'Cortland', 'Empire', 'Golden Delicious', 'McIntosh' and 'Northern Spy'. Internal symptoms 'Braeburn', 'Cripps Pink', 'Fuji', 'Honeycrisp' and 'Pacific Rose'	Elevated CO_2 concentrations during storage environment	Maintain optimum CO_2 levels during storage, DPA inhibits CO_2 injury, DCA storage
Leather blotch	Rough round/oval areas with, the colour turned brown with an irregular border, corking in the cortex, lesions appear 1 to 4 months after harvest	Braeburn', 'Cameo', 'Gala,' 'Fuji', 'Golden Delicious', 'Granny Smith', 'Honeycrisp' and 'Red Delicious'	Susceptibility is high in fruit bitter pit or lenticel blotch-affected fruit, 1-MCP treatment aggravated symptoms	Management measure of bitter pit, avoid using 1-MCP in susceptible cultivars
Low oxygen injury	Skin darkening or purpling, off-flavour development	Any cultivar kept under anaerobic conditions for long durations	Improper gas management, inappropriate airflow during storage	Storage above critically low O_2 limits, DCA storage

Senescent flesh browning/ breakdown	Tissue breakdown, cortex browning, dry and mealy tissues, fruit cracking at later stages, radial flesh browning	'Ambrosia', 'Aroma', 'Baldwin', 'Cox's Orange Pippin', 'Cripps Pink', 'Empire', 'Gala', 'Jonathan', 'Honeycrisp', 'Northern Spy', 'Rome Beauty', 'Red Delicious', 'Rhode Island Greening' and 'Yellow Newtown'	Late harvesting, delayed pre-cooling, high storage temperature, prolonged storage and over-mature fruit with the low calcium content	Harvest at the right maturity, management of storage temperature and pre-harvest nutrient management
Low-temperature breakdown	Diffuse outer cortex browning, core browning in very severe cases	'Braeburn', 'Cox's Orange Pippin', 'Granny Smith', 'Red Delicious', 'Honeycrisp', 'McIntosh', 'Jonathan' and 'Bramley's Seedling'	Longer low-temperature storage, low calcium and phosphorus content, high relative humidity and CO_2 in the storage atmosphere	Proper temperature management, pre-harvest nutrient management
Chilling injury/stem end flesh browning	Diffuse skin browning often originates from the stem end, fruit kept at ambient conditions after low-temperature storage	'Empire' and 'Gala'	Storage temperatures below safe limits	Proper storage temperature management
Soggy scald/breakdown	Brown lesions on skin and flesh, first 6–8 weeks of storage and is not progressive after removal to warm temperatures	'Ellison'Orange', 'Fuji', 'Golden Delicious', 'Honeycrisp', 'Jonathan', 'McIntosh', 'Red Delicious', 'Rome Beauty' and 'Scifresh' (Jazz™)	Over maturity, low-temperature storage < 2–3°C, wet conditions before harvest	Harvest at the right maturity, management of storage temperatures
Superficial scald/storage scald	Brown or black patches on the skin surface, light discolouration of the skin, spotty browning associated with lenticels, sunken areas on the skin, death of hypodermal tissues	'Cortland', 'Fuji', 'Granny Smith' and 'Red Delicious'	Long-term storage, accumulation of α-farnesene in the peel over extended storage periods	Proper storage temperature management, DPA application, DCA storage

(*Continued*)

Table 4 (*Continued*)

Disorder	Symptoms	Susceptible Cultivars	Causes	Control
Watercore	Translucent water-soaked appearance associated with vascular bundles core	Red Delicious', 'Jonathan', 'Fuji', 'Gloster', 'Granny Smith' and 'Orin'	High potassium and magnesium, low calcium content, watercored fruit stored in CA with elevated CO_2 levels increases the risk of breakdown	Harvest at the right maturity, storage temperature management

Source: Watkins and Mattheis (2019).1-MCP = 1-methylcyclopropene; CA = controlled atmosphere; DCA = dynamic controlled atmosphere; DPA = diphenylamine.

fruit quality. The susceptibility to certain disorders varies among cultivars, geographic locations, storage conditions, irrigation and nutrition management of the tree, stage of harvest and postharvest practices. For about more than four decades, physiological disorders such as superficial scald were controlled by the application of antioxidant DPA, usually along with CA conditions (Dias et al., 2020). But the usage of DPA is now being restricted in several countries, as it could be a potential hazard to aquatic environments and carcinogenic to some mammals (Drzyzga, 2003). Usage of ethylene antagonists such as 1-MCP and CA during storage has been proved to be one of the successful strategies to control postharvest disorders and maintain fruit quality (Dias et al., 2020). Apart from maintenance of an ideal storage environment and nutrient status, several other alternatives such as DCA, intermittent warming, edible coatings and ethylene management can be used to reduce the incidence of postharvest disorders. Recently, Torres (2020) discovered that squalane from olive oil (0.1–100% v/v) is effective in controlling apple surface scald on apple fruit (>98%) following 180 days of storage. Some of the major postharvest physiological disorders in apple fruit are summarized in Table 4.

4 Packaging

The type of packages used for apples is based on the convenient size of the lot used for distribution and marketing. The overall purpose of the packaging is to protect the produce against any mechanical damage, maintain the quality, reduce moisture loss and ease of handling. A good package for fruit marketing always has traceability details and an attractive design for consumers. Ideal packaging material is to be selected while considering the produce behaviour, shelf-life, type of storage used, handling and transport methods, merchandising, consumer attitude and recyclability (Gustafsson et al., 2006; Han et al., 2018).

In general, after harvest, the apple fruit is kept in plastic/wooden bins (Fig. 2) and transported to the packhouse for pre-cooling and packaging operations. Corrugated boxes or plastic crates are used to transport the produce to the retail/wholesale market (Figs 3 and 4). Fruit is arranged on a moulded soft board tray to prevent any possible impact bruises and rupturing during transit.

With advancements in polymer engineering, the usage of polymer films has exponentially increased in postharvest handling of the apple fruit. The use can range from the pallet or bin covers, box polylines, consumer bagging or thin wrapping (Gul et al., 2018). MAP technology modifies the gas composition such as O_2, CO_2 and nitrogen inside the package to prolong the shelf-life of the packed produce. The modification happens because of the continued respiration of fruit inside the package increasing levels of CO_2 and lowering the O_2 levels. The O_2 and CO_2 concentrations in the package depend upon the rates of respiration, the mass of the produce, the permeability of the films to

Figure 2 Apple fruit is kept in wooden bins following harvest.

Figure 3 Corrugated boxes with moulded soft board trays.

Figure 4 Apple fruit arranged in foldable plastic crates with moulded soft board trays.

different gases and the surface area and thickness of the film. The apple fruit packed in the MAP exhibited significantly higher fruit firmness, lower moisture and weight loss when compared to the normal atmosphere-stored fruit (Rocha et al., 2004). The fruit quality of Brazilian 'Eva' apples was maintained for up to 7 months packed in MAP (Fante et al., 2014). The fruit packed in high-density polyethene film exhibited the lowest ethylene production, possibly due to the increased CO_2 concentrations inside the package (de Wild et al., 1999).

5 Transportation

The temperature management and protection of fruit from mechanical or impact injuries are important factors for maximum transit life. However, managing these variables would be meaningful only if the produce being transported is of optimum quality. Apples should be transported at the pre-climacteric stage with peel exhibiting lightening of the ground colour. Proper postharvest handling including sorting, cleaning, pre-cooling, temperature management and packaging, directly affects the quality of produce during and after transportation. For long-distance marketing, it is recommended to load only the same type of products as much as possible. Fruit absorbs aroma volatiles of commodities such as potatoes, garlic, onion, meat and cheese. It is strongly recommended to avoid mixing the apple loads with commodities that are high ethylene releasing or ethylene sensitive (Gross et al., 2016). The transport container should be fully loaded as temperature control is difficult in partially loaded. The pallets should be arranged in a pattern that allows optimum air circulation through cargo (Kerbel, 2021).

The reefer containers are not generally constructed to immediately cool the fruit to bring down the pulp temperature and maintain the temperature of the produce already cooled. Any warm produce loaded in the container will take several days for pulp temperature to reach recommended temperature settings, significantly reducing the transit life of the fruit. The CA system in the transport containers can also be used for apples, opening new market opportunities around the globe. CA can uphold the fruit quality for a relatively prolonged duration without the use of chemicals or fungicides during transit when sailing is slower and on extended routes with multiple stops (Kerbel, 2021).

For domestic, on-road transportation, the trucks used should be properly serviced and equipped with necessary shock-absorbers to prevent any impact and pressure injuries. The fruit cargo must be covered appropriately to protect them from external environmental factors such as rain, winds and sun (Fig. 5). For non-refrigerated trucks, it is recommended to transport in a cooler part of the day or night (Gross et al., 2016).

Figure 5 Apple fruit is ready for domestic truck transport with proper covering.

6 Conclusion and future trends in research

Within the group of climacteric fruits, apples have a longer storage life. The main issues during postharvest handling are loss of fruit weight and firmness, development of physiological internal and external disorders and quality losses. The maximum storage duration while maintaining optimum fruit quality to facilitate international trade of apple fruit will be strengthened through a holistic approach encompassing preharvest production technology to proper produce handling, downregulating physiological processes and management of diseases and disorders. There are various technologies for storage, ethylene management and quality retention available and are being used by different stakeholders in the apple industry. Recently, DCA storage technology is gaining great interest among apple industry stakeholders to prolong the storage life and uphold the apple fruit quality. The physiology of various cultivars of apple is different, and the effectiveness of the treatments varies among cultivars. The protocols for postharvest handling of some popular cultivars are established to a maximum extent, but still there exists a scope to identify the cultivar-specific postharvest issues and design respective solutions.

7 Where to look for further information

The introduction to the basic concepts of postharvest fruit physiology and different technologies affecting the fruit behaviour from the instance of the harvest till it reaches the consumer can be referred from *Postharvest: An Introduction to the Physiology and Handling of Fruit and Vegetables* by Wills and Golding (see Wills and Golding, 2016).

Ethylene management in apples has been found to influence several postharvest changes directly or indirectly in the fruit. The technologies to manipulate storage environments as well as antagonize ethylene action in apple fruit have revolutionized the global apple industry by increasing their capacity to be stored for a long duration without any significant changes in the quality. As mentioned in Section 6, there is still a large scope to research on the methods to mitigate the quality losses during postharvest storage. Our work on using different innovations to maintain the postharvest quality of apple fruit can be referred from the study by Tokala (2019), (2020a,b), (2021a, b). Management of erratic incidence of internal flesh browning and lenticel breakdown, superficial scald, other postharvest physiological disorders (included in Section 3.5) and postharvest quality management of organically grown apples warrants to be investigated and can be referred from the study by Torres (2020) and Watkins and Mattheis (2019).

8 References

Abeles, F. B., Morgan, P. W. and Saltveit, M. E., Jr. (2012). *Ethylene in Plant Biology*. Academic Press Inc, Cambridge.

Abi Tarabay, P., Chahine-Tsouvalakis, H., Tawk, S. T., Nemer, N. and Habib, W. (2018). Reduction of food losses in Lebanese apple through good harvesting and postharvest practices. *Annals of Agricultural Sciences* 63(2), 207–213.

AgroFresh. (2022). We grow confidence. Available at: https://www.agrofresh.com/technologies (accessed March 1, 2022).

Ali, U., Kanwar, S., Yadav, K., Basu, S. and Mazumder, K. (2019). Effect of arabinoxylan and β-glucan stearic acid ester coatings on post-harvest quality of apple (Royal delicious). *Carbohydrate Polymers* 209, 338–349.

Amarante, C. V. Td, Silveira, J. P. G., Freitas, S. Td, Steffens, C. A. and Mitcham, E. J. (2021). Fruit quality of 'Braeburn' apple trees sprayed at post-bloom and preharvest with prohexadione-calcium and GA_{4+7}. *Revista Brasileira de Fruticultura* 43(1), 1-6.

Ambaw, A., Mukama, M. and Opara, U. L. (2017). Analysis of the effects of package design on the rate and uniformity of cooling of stacked pomegranates: Numerical and experimental studies. *Computers and Electronics in Agriculture* 136, 13–24.

Apelbaum, A., Sisler, E. C., Feng, X. and Goren, R. (2008). Assessment of the potency of 1-substituted cyclopropenes to counteract ethylene-induced processes in plants. *Plant Growth Regulation* 55(2), 101–113.

Arts, J. H., Hadi, M., Irfan, M. A., Keene, A. M., Kreiling, R., Lyon, D., Maier, M., Michel, K., Petry, T., Sauer, U. G. and Warheit, D. (2015). A decision-making framework for the grouping and testing of nanomaterials (DF4nanoGrouping). *Regulatory Toxicology and Pharmacology* 71(2), 1-27.

Baier, M., Ehlbeck, J., Knorr, D., Herppich, W. B. and Schlüter, O. (2015). Impact of plasma processed air (PPA) on quality parameters of fresh produce. *Postharvest Biology and Technology* 100, 120–126.

Baldwin, E. A., Hagenmaier, R. and Bai, J. (2011). *Edible Coatings and Films to Improve Food Quality*. CRC Press, Boca Raton, FL, pp. 1-12.

Baritelle, A. L., Hyde, G. M., Fellman, J. K. and Varith, J. (2001). Using 1-MCP to inhibit the influence of ripening on impact properties of pear and apple tissue. *Postharvest Biology and Technology* 23(2), 153–160.

Barry, C. S., Llop-Tous, M. I. and Grierson, D. (2000). The regulation of 1-aminocyclopropane-1-carboxylic acid synthase gene expression during the transition from system-1 to system-2 ethylene synthesis in tomato. *Plant Physiology* 123(3), 979–986.

Baswal, A. K. and Ramezanian, A. (2021). 1-Methylcyclopropene potentials in maintaining the postharvest quality of fruits, vegetables, and ornamentals: A review. *Journal of Food Processing and Preservation* 45(1), 15129.

Bazarova, V. I. (1982). Use of ozone in storage of apples. *Food Science and Technology Abstracts* 14(11), J1653.

Blanke, M. M. (2014). Reducing ethylene levels along the food supply chain: A key to reducing food waste? *Journal of the Science of Food and Agriculture* 94(12), 2357–2361.

Blankenship, S. M. and Dole, J. M. (2003). 1-Methylcyclopropene: A review. *Postharvest Biology and Technology* 28(1), 1–25.

Blanpied, G. D., Smock, R. M. and Kollas, D. A. (1967). Effect of Alar on optimum harvest dates and keeping quality of apples. *Proceedings of the American Society of Horticultural Sciences* 90, 467–474.

Bridgemohan, P., Mohammed, M. and Tokala, V. Y. (2021). Cooling requirements of selected perishable crops during storage. In: Tokala, V. Y. and Mohammed, M. (Eds.), *Cold Chain Management for the Fresh Produce Industry in the Developing World*. CRC Press, Boca Raton, FL, pp. 19–30.

Burg, S. P. and Burg, E. A. (1962). Role of ethylene in fruit ripening. *Plant Physiology* 37(2), 179–189.

Chen, Y. Q., Cheng, J. H. and Sun, D. W. (2020). Chemical, physical and physiological quality attributes of fruit and vegetables induced by cold plasma treatment: Mechanisms and application advances. *Critical Reviews in Food Science and Nutrition* 60(16), 2676–2690.

Concha-Meyer, A., Eifert, J. D., Williams, R. C., Marcy, J. E. and Welbaum, G. E. (2015). Shelf life determination of fresh blueberries (*Vaccinium corymbosum*) stored under controlled atmosphere and ozone. *International Journal of Food Science* 2015, 164143.

Darras, A. I., Tsikaloudakis, G., Lycoskoufis, I., Dimitriadis, C. and Karamousantas, D. (2020). Low doses of UV-C irradiation affects growth, fruit yield and photosynthetic activity of tomato plants. *Scientia Horticulturae* 267, 109357.

Dauny, P. T. and Joyce, D. C. (2002). 1-MCP improves storability of 'Queen Cox' and 'Bramley' apple fruit. *Hortscience* 37(7), 1082–1085.

de Wild, H. P. J., Woltering, E. J. and Peppelenbos, H. W. (1999). Carbon dioxide and 1-MCP inhibit ethylene production and respiration of pear fruit by different mechanisms. *Journal of Experimental Botany* 50(335), 837–844.

Delele, M. A., Tijskens, E., Atalay, Y. T., Ho, Q. T., Ramon, H., Nicolaï, B. M. and Verboven, P. (2008). Combined discrete element and CFD modelling of airflow through random stacking of horticultural products in vented boxes. *Journal of Food Engineering* 89(1), 33–41.

DeLong, J. M., Prange, R. K. and Harrison, P. A. (2004). The influence of 1-methylcyclopropene on 'Cortland' and 'McIntosh' apple quality following long-term storage. *HortScience* 39(5), 1062–1065.

Dhall, R. K. (2013). Advances in edible coatings for fresh fruits and vegetables: A review. *Critical Reviews in Food Science and Nutrition* 53(5), 435–450.

Dias, C., L Amaro, A., Salvador, Â. C., Silvestre, A. J. D., Rocha, S. M., Isidoro, N. and Pintado, M. (2020). Strategies to preserve postharvest quality of horticultural crops and superficial scald control: From diphenylamine antioxidant usage to more recent approaches. *Antioxidants* 9(4), 356.

Domínguez, I., Lafuente, M. T., Hernández-Muñoz, P. and Gavara, R. (2016). Influence of modified atmosphere and ethylene levels on quality attributes of fresh tomatoes (*Lycopersicon esculentum* Mill.). *Food Chemistry* 209, 211–219.

Drzyzga, O. (2003). Diphenylamine and derivatives in the environment: a review. Chemosphere 53(8), 809–818.

Duan, Y., Wang, G. B., Fawole, O. A., Verboven, P., Zhang, X. R., Wu, D., Opara, U. L., Nicolai, B. and Chen, K. (2020). Postharvest precooling of fruit and vegetables: A review. *Trends in Food Science and Technology* 100, 278–291.

Ekezie, F. G. C., Sun, D. W. and Cheng, J. H. (2017). A review on recent advances in cold plasma technology for the food industry: Current applications and future trends. *Trends in Food Science and Technology* 69, 46–58.

El-Anany, A. M., Hassan, G. F. A. and Ali, F. R. (2009). Effects of edible coatings on the shelf-life and quality of Anna apple (*Malus domestica* Borkh) during cold storage. *Journal of Food Technology* 7(1), 5–11.

Escribano, S., Sugimoto, N., Macnish, A. J., Biasi, W. V. and Mitcham, E. J. (2017). Efficacy of liquid 1-methylcyclopropene to delay ripening of 'Bartlett' pears. *Postharvest Biology and Technology* 126, 57–66.

Esua, O. J., Chin, N. L., Yusof, Y. A. and Sukor, R. (2020). A review on individual and combination technologies of UV-C radiation and ultrasound in postharvest handling of fruits and vegetables. *Processes* 8(11), 1433.

Fan, X., Mattheis, J. P. and Fellman, J. K. (1998). A role for jasmonates in climacteric fruit ripening. *Planta* 204(4), 444–449.

Fanourakis, D., Pieruschka, R., Savvides, A., Macnish, A. J., Sarlikioti, V. and Woltering, E. J. (2013). Sources of vase life variation in cut roses: A review. *Postharvest Biology and Technology* 78, 1–15.

Fante, C. A., Boas, A. C. V., Paiva, V. A., Pires, C. R. F. and Lima, L. CdO. (2014). Modified atmosphere efficiency in the quality maintenance of Eva apples. *Food Science and Technology* 34(2), 309–314.

Farina, V., Passafiume, R., Tinebra, I., Palazzolo, E. and Sortino, G. (2020). Use of aloe vera gel-based edible coating with natural anti-browning and anti-oxidant additives to improve post-harvest quality of fresh-cut 'Fuji' apple. *Agronomy* 10(4), 515.

Funt, R. C., Cameron, E. A. and Banks, N. H. (1999). The effect of apple fruit bruising on total returns. Research circular. *Ohio Agricultural Research and Development Center* 299, 55–61.

Gardesh, A. S. K., Badii, F., Hashemi, M., Ardakani, A. Y., Maftoonazad, N. and Gorji, A. M. (2016). Effect of nano chitosan-based coating on climacteric behaviour and postharvest shelf-life extension of apple cv. Golab Kohanz. *LWT* 70, 33–40.

Gianfagna, T. J., Durner, E. F. and Salas, S. Q. (1989). Evaluation of plant growth regulators as possible replacement compounds for alar in apple production. In: *Acta Horticulturae Symposium on Growth Regulators in Fruit Production* (Vol. 239), pp. 439–442.

Gillies, J. Z., Gillies, C. W., Suenram, R. D. and Lovas, F. J. (1988). The ozonolysis of ethylene. microwave spectrum, molecular structure, and dipole moment of ethylene primary ozonide (1, 2, 3-trioxolane). *Journal of the American Chemical Society* 110(24), 7991–7999.

Gross, K. C., Wang, C. Y. and Saltveit, M. E. (Eds.). (2016). *USDA Handbook 66. The Commercial Storage of Fruits, Vegetables, and Florist and Nursery Stocks*, p. 792.

Guerreiro, A. C., Gago, C. M. L., Faleiro, M. L., Miguel, M. G. C. and Antunes, M. D. C. (2017). The effect of edible coatings on the nutritional quality of 'Bravo de Esmolfe' fresh-cut apple through shelf-life. *LWT* 75, 210–219.

Gul, K., Mir, N. A., Singh, P. and Wani, A. A. (2018). Postharvest biology and technology of apple. In: Mir, S. A., Shah, M. A. and Mir, M. M. (Eds.), *Postharvest Biology and Technology of Temperate Fruits*. Springer International Publishing, Switzerland, AG, pp. 223–244.

Gustafsson, K., Joönson, G., Smith, D. and Sparks, L. (2006). Packaging and fresh food. In: *Retailing Logistics & Fresh Food Packaging: Managing Change in the Supply Chain*. The Chartered Institute of Logistics and Transport, UK, Kogan Page, London, UK, pp. 69–104.

Hadian-Deljou, M., Esna-Ashari, M. and Sarikhani, H. (2017). Effect of pre-and post-harvest salicylic acid treatments on quality and antioxidant properties of 'Red Delicious' apples during cold storage. *Advances in Horticultural Science* 31(1), 31–38.

Han, J. W., Ruiz-Garcia, L., Qian, J. P. and Yang, X. T. (2018). Food packaging: A comprehensive review and future trends. *Comprehensive Reviews in Food Science and Food Safety* 17(4), 860–877.

Hazel Technology Inc. (2022). Available at: https://www.hazeltechnologies.com/products (accessed March 1, 2022).

Hemmaty, S., Moallemi, N. and Naseri, L. (2006). Shelf-life and quality of apple fruits in response to postharvest application of UV-C radiation. *Journal of Applied Horticulture* 8(2), 114–116.

Horvitz, S. and Cantalejo, M. J. (2014). Application of ozone for the postharvest treatment of fruits and vegetables. *Critical Reviews in Food Science and Nutrition* 54(3), 312–339.

Iqbal, N., Khan, N. A., Ferrante, A., Trivellini, A., Francini, A. and Khan, M. I. R. (2017). Ethylene role in plant growth, development and senescence: Interaction with other phytohormones. *Frontiers in Plant Science* 8, 475.

Isidoro, N. and Almeida, D. P. F. (2006). α-Farnesene, conjugated trienols, and superficial scald in 'Rocha' pear as affected by 1-methylcyclopropene and diphenylamine. *Postharvest Biology and Technology* 42(1), 49–56.

Jahanshahi, B., Jafari, A., Vazifeshenas, M. R. and Gholamnejad, J. (2018). A novel edible coating for apple fruits. *Journal of Horticulture and Postharvest Research* 1, 63–72.

Johnson, D. S. and Ertan, U. (1983). Interaction of temperature and oxygen level on the respiration rate and storage quality of Idared apples. *Journal of Horticultural Science* 58(4), 527–533.

Johnston, J. W., Hewett, E. W. and Hertog, M. L. A. T. M. (2005). Apple (*Malus domestica*) softening in the postharvest cool chain: Effects of delayed cooling and shelf-life temperatures. *New Zealand Journal of Crop and Horticultural Science* 33(3), 283–292.

Jongsri, P., Wangsomboondee, T., Rojsitthisak, P. and Seraypheap, K. (2016). Effect of molecular weights of chitosan coating on postharvest quality and physicochemical characteristics of mango fruit. *LWT* 73, 28–36.

Juhnevica-Radenkova, K., Radenkovs, V., Kundzins, K. and Seglina, D. (2019). Effect of ozone treatment on the microstructure, chemical composition and sensory quality of apple fruits. *Food Science and Technology International* 25(3), 252–267.

Jung, S. K. and Watkins, C. B. (2008). Superficial scald control after delayed treatment of apple fruit with diphenylamine (DPA) and 1-methylcyclopropene (1-MCP). *Postharvest Biology and Technology* 50(1), 45–52.

Kader, A. A. and Rolle, R. S. (2004). Postharvest treatments designed to manipulate the environment around produce in order to enhance quality. In: *The Role of Post-Harvest Management in Assuring the Quality and Safety of Horticultural Produce* (Vol. 152). Food and Agriculture Organisation, Rome, pp. 35–43.

Kadzińska, J., Bryś, J., Ostrowska-Ligęza, E., Estéve, M. and Janowicz, M. (2020). Influence of vegetable oils addition on the selected physical properties of apple-sodium alginate edible films. *Polymer Bulletin* 77(2), 883–900.

Keller, N., Ducamp, M. N., Robert, D. and Keller, V. (2013). Ethylene removal and fresh product storage: A challenge at the frontiers of chemistry. Toward an approach by photocatalytic oxidation. *Chemical Reviews* 113(7), 5029–5070.

Kerbel, E. (2021). Refrigerated transport in marine containers and cold transport logistics. In: Tokala, V. Y. and Mohammed, M. (Eds.), *Cold Chain Management for the Fresh Produce Industry in the Developing World*. CRC Press, Boca Raton, FL, pp. 99–118.

Khalifa, I., Barakat, H., El-Mansy, H. A. and Soliman, S. A. (2017). Preserving apple (Malus domestica var. Anna) fruit bioactive substances using olive wastes extract-chitosan film coating. *Information Processing in Agriculture* 4(1), 90–99.

Khan, N., Riaz, A., Rahman, Z., Mawa, J. U. and Begum, H. (2019). Shelf-life assessment of apple fruit coated with aloe vera gel and calcium chloride. *Pure and Applied Biology (PAB)* 8(3), 1876–1889.

Kingston, C. M. (1992). Maturity indices for apple and pear. *Horticultural Reviews* 13(32), 402.

Kitinoja, L. and Kader, A. A. (2002). *Small-Scale Postharvest Handling Practices: A Manual for Horticultural Crops*. University of California, Postharvest Technology Research and Information Center, Davis, CA.

Kramer, G. F., Wang, C. Y. and Conway, W. S. (1991). Inhibition of softening by polyamine application in 'Golden Delicious' and 'McIntosh' apples. *Journal of the American Society for Horticultural Science* 116(5), 813–817.

Larrigaudière, C., Vilaplana, R., Valentines, M. C., Soria, Y. and Recasens, I. (2004). Effects of 1-methylcyclopropene on antioxidant metabolism in cold-stored fruit. In: *V International Postharvest Symposium 682*, pp. 957–964.

Li, L., Lichter, A., Chalupowicz, D., Gamrasni, D., Goldberg, T., Nerya, O., Ben-Arie, R. and Porat, R. (2016). Effects of the ethylene-action inhibitor 1-methylcyclopropene on postharvest quality of non-climacteric fruit crops. *Postharvest Biology and Technology* 111, 322–329.

Liew, C. L. and Prange, R. K. (1994). Effect of ozone and storage temperature on postharvest diseases and physiology of carrots (*Daucus carota* L.). *Journal of the American Society for Horticultural Science* 119(3), 563–567.

Logfresh – Shandong Aoweite Biotechnology Co. Ltd. (2022). Available at: https://www.chinesepost-harvest.com/1-methylcyclopropene (accessed March 1, 2022).

Looney, N. E. (2018). Growth regulator usage in apple and pear production. In: Nickell, L. G. (Ed.), *Plant Growth Regulating Chemicals*. CRC Press, Boca Raton, FL, pp. 1–26.

Lu, C., Ding, J., Park, H. K. and Feng, H. (2020). High-intensity ultrasound as a physical elicitor affects secondary metabolites and antioxidant capacity of tomato fruits. *Food Control* 113, 107176.

Lufu, R., Ambaw, A. and Opara, U. L. (2019). The contribution of transpiration and respiration processes in the mass loss of pomegranate fruit (cv. Wonderful). *Postharvest Biology and Technology* 157, 110982.

Luo, H., Song, J., Toivonen, P., Gong, Y., Forney, C., Palmer, L. C., Fillmore, S., Pang, X. and Zhang, Z. (2018). Proteomic changes in 'Ambrosia' apple fruit during cold storage and in response to delayed cooling treatment. *Postharvest Biology and Technology* 137, 66–76.

Lurie, S. and Watkins, C. B. (2012). Superficial scald, its etiology and control. *Postharvest Biology and Technology* 65, 44–60.

MacLean, D. D., Murr, D. P., DeEll, J. R. and Horvath, C. R. (2006). Postharvest variation in apple (*Malus × domestica* Borkh.) flavonoids following harvest, storage, and 1-MCP treatment. *Journal of Agricultural and Food Chemistry* 54(3), 870–878.

Mahajan, P. V., Caleb, O. J., Singh, Z., Watkins, C. B. and Geyer, M. (2014). Postharvest treatments of fresh produce. *Philosophical Transactions. Series A, Mathematical, Physical, and Engineering Sciences* 372(2017), 20130309.

Maletsika, P., Nanos, G. D., Papoulia, I. and Vasilakakis, M. (2014). The effect of postharvest application of O_3 and 1-MCP on 'Red Chief' apple quality during prolonged air cold storage. *Acta Horticulturae* (1079), 429–434.

Manousaridis, G., Nerantzaki, A., Paleologos, E. K., Tsiotsias, A., Savvaidis, I. N. and Kontominas, M. G. (2005). Effect of ozone on microbial, chemical and sensory attributes of shucked mussels. *Food Microbiology* 22(1), 1–9.

Maringgal, B., Hashim, N., Mohamed Amin Tawakkal, I. S. and Muda Mohamed, M. T. (2020). Recent advance in edible coating and its effect on fresh/fresh-cut fruits quality. *Trends in Food Science and Technology* 96, 253–267.

Martínez-Romero, D., Bailén, G., Serrano, M., Guillén, F., Valverde, J. M., Zapata, P., Castillo, S. and Valero, D. (2007). Tools to maintain postharvest fruit and vegetable quality through the inhibition of ethylene action: A review. *Critical Reviews in Food Science and Nutrition* 47(6), 543–560.

Maryam, A., Anwar, R., Malik, A. U., Raheem, M. I. U., Khan, A. S., Hasan, M. U., Hussain, Z. and Siddique, Z. (2021). Combined aqueous ozone and ultrasound application inhibits microbial spoilage, reduces pesticide residues and maintains storage quality of strawberry fruits. *Journal of Food Measurement and Characterization* 15(2), 1437–1451.

Masia, A. (1998). Superoxide dismutase and catalase activities in apple fruit during ripening and post-harvest and with special reference to ethylene. *Physiologia Plantarum* 104(4), 668–672.

Masia, A., Ventura, M., Gemma, H. and Sansavini, S. (1998). Effect of some plant growth regulator treatments on apple fruit ripening. *Plant Growth Regulation* 25(2), 127–134.

Mditshwa, A., Fawole, O. A. and Opara, U. L. (2018). Recent developments on dynamic controlled atmosphere storage of apples - A review. *Food Packaging and Shelf Life* 16, 59–68.

Mercier, S., Villeneuve, S., Mondor, M. and Uysal, I. (2017). Time-temperature management along the food cold chain: A review of recent developments. *Comprehensive Reviews in Food Science and Food Safety* 16(4), 647–667.

Mir, N. A., Curell, E., Khan, N., Whitaker, M. and Beaudry, R. M. (2001). Harvest maturity, storage temperature, and 1-MCP application frequency alter firmness retention and chlorophyll fluorescence of 'Redchief Delicious' apples. *Journal of the American Society for Horticultural Science* 126(5), 618–624.

Mitchell, G. (1992). *Postharvest Technology of Horticultural Crops*. Edited by A. Kader. The University of California, Division of Agriculture and Natural Resources Publication, p. 3311.

Muhammad, A. I., Xiang, Q., Liao, X., Liu, D. and Ding, T. (2018). Understanding the impact of non-thermal plasma on food constituents and microstructure—A review. *Food and Bioprocess Technology* 11(3), 463–486.

Musacchi, S. and Serra, S. (2018). Apple fruit quality: Overview on pre-harvest factors. *Scientia Horticulturae* 234, 409–430.

Oliveira, J., Gloria, E. M., Parisi, M. C. M., Baggio, J. S., Silva, P. P. M., Ambrosio, C. M. S. and Spoto, M. H. F. (2019). Antifungal activity of essential oils associated with carboxymethylcellulose against *Colletotrichum acutatum* in strawberries. *Scientia Horticulturae* 243, 261–267.

Onik, J. C., Wai, S. C., Li, A., Lin, Q., Sun, Q., Wang, Z. and Duan, Y. (2021). Melatonin treatment reduces ethylene production and maintains fruit quality in apple during postharvest storage. *Food Chemistry* 337, 127753.

Onik, J. C., Xie, Y., Duan, Y., Hu, X., Wang, Z. and Lin, Q. (2019). UV-C treatment promotes quality of early ripening apple fruit by regulating malate metabolizing genes during postharvest storage. *PLoS ONE* 14(4), e0215472.

Palou, L., Smilanick, J. L., Crisosto, C. H. and Mansour, M. (2001). Effect of gaseous ozone exposure on the development of green and blue molds on cold stored citrus fruit. *Plant Disease* 85(6), 632–638.

Paull, R. (1999). Effect of temperature and relative humidity on fresh commodity quality. *Postharvest Biology and Technology* 15(3), 263–277.

Paulson, G. S., Hull, L. A. and Biddinger, D. J. (2005). Effect of a plant growth regulator prohexadione-calcium on insect pests of apple and pear. *Journal of Economic Entomology* 98(2), 423–431.

Prange, R. K., Delong, J. M., Harrison, P., McLean, S., Scrutton, J. and Cullen, J. (2007). *Method and Apparatus for Monitoring a Condition in Chlorophyll Containing Matter*. U.S. Patent, No. WO/2002/006795.

Rocha, A. M. C. N., Barreiro, M. G. and Morais, A. M. M. B. (2004). Modified atmosphere package for apple 'Bravo de Esmolfe'. *Food Control* 15(1), 61–64.

Rudell, D. R., Fellman, J. K. and Mattheis, J. P. (2005). Preharvest application of methyl jasmonate to 'Fuji' apples enhances red coloration and affects fruit size, splitting, and bitter pit incidence. *HortScience* 40(6), 1760–1762.

Saltveit, M. E. (1999). Effect of ethylene on quality of fresh fruits and vegetables. *Postharvest Biology and Technology* 15(3), 279–292.

Saltveit, M. E. (2003). Is it possible to find an optimal controlled atmosphere? *Postharvest Biology and Technology* 27(1), 3–13.

Schmitz, M. and Noga, G. (1996). α-tocopherol reduced environmental stress and improved fruit quality. In: *II Workshop on Pome Fruit* (Vol. 466), pp. 89–94.

Schnabel, U., Niquet, R., Schlüter, O., Gniffke, H. and Ehlbeck, J. (2015). Decontamination and sensory properties of microbiologically contaminated fresh fruits and vegetables by microwave plasma processed air (PPA). *Journal of Food Processing and Preservation* 39(6), 653–662.

Schotsmans, W. C., Prange, R. K. and Binder, B. M. (2009). 1-Methylcyclopropene: Mode of action and relevance in postharvest horticulture research. *Horticultural Reviews* 35, 263–313.

Shafiq, M., Singh, Z. and Khan, A. S. (2011a). Delayed harvest and cold storage period influence ethylene production, fruit firmness and quality of 'Cripps Pink' apple. *International Journal of Food Science and Technology* 46(12), 2520–2529.

Shafiq, M., Singh, Z. and Khan, A. S. (2011b). Delayed harvest improves red blush development and quality of 'Cripps Pink' apple. *Scientia Horticulturae* 129(4), 715–723.

Shafiq, M., Singh, Z. and Khan, A. S. (2011c). Pre-harvest spray application of methyl jasmonate improves red blush and flavonoid content in 'Cripps Pink' apple. *The Journal of Horticultural Science and Biotechnology* 86(4), 422–430.

Shafiq, M., Singh, Z. and Khan, A. S. (2013). Time of methyl jasmonate application influences the development of 'Cripps Pink' apple fruit colour. *Journal of the Science of Food and Agriculture* 93(3), 611–618.

Shen, X., Su, Y., Hua, Z., Sheng, L., Mendoza, M., He, Y., Green, T., Hanrahan, I., Blakey, R. and Zhu, M. J. (2021). Effectiveness of low-dose continuous gaseous ozone in controlling *Listeria innocua* on red delicious apples during 9-month commercial cold storage. *Frontiers in Microbiology* 12, 712757.

Singh, D. P. and Packirisamy, G. (2022). Biopolymer based edible coating for enhancing the shelf life of horticulture products. *Food Chemistry. Molecular Sciences* 4, 100085.

Singh, Z., Payne, A. D., Khan, S. A. K. U. and Musa, M. M. (2018). Method of retarding an ethylene response. U.S. Patent Application No. 15/772, 324.

Sisler, E. C. (2006). The discovery and development of compounds counteracting ethylene at the receptor level. *Biotechnology Advances* 24(4), 357–367.

Sisler, E. C., Grichko, V. P. and Serek, M. (2006). Interaction of ethylene and other compounds with the ethylene receptor: Agonists and antagonists. In: Khan, N. A. (Ed.), *Ethylene Action in Plants*. Springer, Berlin, Heidelberg, pp. 1–34.

Skog, C. L. and Chu, L. J. (2001). Effect of ozone on qualities of fruits and vegetables in cold storage. *Canadian Journal of Plant Science* 81(4), 773–778.

Song, H. Y., Jo, W. S., Song, N. B., Min, S. C. and Song, K. B. (2013). Quality change of apple slices coated with *Aloe vera* gel during storage. *Journal of Food Science* 78(6), C817–C822.

Steffens, C. A., Miqueloto, T., Fernandes, R. C., Demari, C. K. P., Anami, J. M., Lugaresi, A. and Amarante, C. V. Td (2021). Treatment with nitric oxide preserves the quality of cold stored 'Cripps Pink' apples. *Bragantia* 80, 1–9.

Steinite, I., Gailite, A. and Ievinsh, G. (2004). Reactive oxygen and ethylene are involved in the regulation of regurgitant-induced responses in bean plants. *Journal of Plant Physiology* 161(2), 191–196.

Taiz, L., Zeiger, E., Møller, I. M. and Murphy, A. (2015). Ethylene: The gaseous hormone. In: *Plant Physiology and Development*. Sinauer Associates, Sunderland, MA, pp. 519–538.

Tappi, S., Ragni, L., Tylewicz, U., Romani, S., Ramazzina, I. and Rocculi, P. (2019). Browning response of fresh-cut apples of different cultivars to cold gas plasma treatment. *Innovative Food Science & Emerging Technologies* 53, 56–62.

Thakur, R., Pristijono, P., Scarlett, C. J., Bowyer, M., Singh, S. P. and Vuong, Q. V. (2019). Starch-based edible coating formulation: Optimization and its application to improve the postharvest quality of 'Cripps Pink' apple under different temperature regimes. *Food Packaging and Shelf Life* 22, 100409.

Thompson, A. K. (2010). Effects and interactions of CA storage. In: *Controlled Atmosphere Storage of Fruits and Vegetables*. CABI Publishing, UK, pp. 11–26.

Thompson, J. F., Mitchell, F. G. and Kasmire, R. F. (2002). Cooling horticultural commodities. In: Kader, A. A. (Ed.), *Postharvest Technology of Horticultural Crops* (3rd edn., Vol. 3311). The University of California, Division of Agriculture and Natural Resources Publication, USA, pp. 97–112.

Thompson, J. F., Mitchell, F. G., Rumsey, T. R., Kasmire, R. F. and Crisosto, C. C. (2008). *Commercial Cooling of Fruits, Vegetables, and Flowers*. University of California, DANR Publication, USA, 21567.

Toivonen, P. and Lu, L. (2005). Studies on elevated temperature, short-term storage of 'Sunrise' summer apples using 1-MCP to maintain quality. *The Journal of Horticultural Science and Biotechnology* 80(4), 439–466.

Tokala, V. Y. (2019). *Postharvest interventions involving novel ethylene antagonists, ozone and Airofresh® in regulating ethylene and maintaining quality of apple and pear fruits*. Doctoral Dissertation. Curtin University, Western Australia. Available at: http://hdl.handle.net/20.500.11937/77988.

Tokala, V. Y., Mahajan, B. V. C. and Bal, J. S. (2020a). Growth regulators in postharvest life of fruits. In: In, J. S. B. (Ed.), *Plant Growth Regulators in Fruit Science*. Kalyani Publishers, India, pp. 15.1–15.43.

Tokala, V. Y., Singh, Z. and Kyaw, P. N. (2020b). Fumigation and dip treatments with 1*H*-cyclopropabenzene and 1*H*-cyclopropa [*b*] naphthalene suppress ethylene production and maintain fruit quality of cold-stored 'Cripps Pink' apple. *Scientia Horticulturae* 272, 109597.

Tokala, V. Y., Singh, Z. and Kyaw, P. N. (2021a). 1*H*-cyclopropabenzene and 1*H*-cyclopropa [*b*] naphthalene fumigation downregulates ethylene production and maintains fruit quality of controlled atmosphere stored 'Granny Smith' apple. *Postharvest Biology and Technology* 176, 111499.

Tokala, V. Y., Singh, Z. and Kyaw, P. N. (2021b). Postharvest fruit quality of apple influenced by ethylene antagonist fumigation and ozonized cold storage. *Food Chemistry* 341(2), 128293.

Tokala, V. Y., Singh, Z. and Kyaw, P. N. (2022). Postharvest quality of 'Cripps Pink' apple fruit influenced by ethylene antagonists during controlled atmosphere storage with photocatalytic oxidation. *Journal of the Science of Food and Agriculture* 102(11), 4484–4490.

Tokala, V. Y., Singh, Z. and Payne, A. D. (2018). Postharvest uses of ozone application in fresh horticultural produce. In: In, G., Paliyath, J., Subramanian, L.-T., Lim, K. S., Subramanian, Handa A. and Mattoo, A. (Eds.), *Postharvest Biology and Nanotechnology*. John Wiley and Sons Inc., New Jersey, pp. 129–170.

Tolouie, H., Mohammadifar, M. A., Ghomi, H. and Hashemi, M. (2018). Cold atmospheric plasma manipulation of proteins in food systems. *Critical Reviews in Food Science and Nutrition* 58(15), 2583–2597.

Torres, C. A. (2020). Composition for controlling superficial scald in pome fruit. U.S. Patent Application Number 16/490046. Available at: https://uspto.report/patent/app/20200008436.

Tzortzakis, N. and Chrysargyris, A. (2017). Postharvest ozone application for the preservation of fruits and vegetables. *Food Reviews International* 33(3), 270–315.

Valero, D., Martínez-Romero, D. and Serrano, M. (2002). The role of polyamines in the improvement of the shelf life of fruit. *Trends in Food Science and Technology* 13(6-7), 228–234.

Veltman, R. H., Verschoor, J. A. and Van Dugteren, J. H. R. (2003). Dynamic control system (DCS) for apples (*Malus domestica* Borkh. cv 'Elstar'): Optimal quality through storage based on product response. *Postharvest Biology and Technology* 27(1), 79–86.

Wang, C. Y., Conway, W. S., Abbott, J. A., Kramer, G. F. and Sams, C. E. (1993). Postharvest infiltration of polyamines and calcium influences ethylene production and texture changes in 'Golden Delicious' apples. *Journal of the American Society for Horticultural Science* 118(6), 801–806.

Watkins, C. B. (2003). Principles and practices of postharvest handling and stress. In: Ferree, I. and David, C. (Eds.), *Apples: Botany, Production, and Uses*. CABI Publishing, Boston, MA, pp. 585–614.

Watkins, C. B. (2006). The use of 1-methylcyclopropene (1-MCP) on fruits and vegetables. *Biotechnology Advances* 24(4), 389–409.

Watkins, C. B. (2008). Overview of 1-methylcyclopropene trials and uses for edible horticultural crops. *Hortscience* 43(1), 86–94.

Watkins, C. B. (2015). Advances in the use of 1-MCP. In: Wills, R. B. H. and Golding, J. B. (Eds.), *Advances in Postharvest Fruit and Vegetable Technology*. CRC Press, Boca Raton, FL, pp. 117–146.

Watkins, C. B. and Mattheis, J. P. (2019). Apple. In: de Freitas, S. T. and Pareek, S. (Eds.), *Postharvest Physiological Disorders in Fruits and Vegetables*. CRC Press, Boca Raton, FL, pp. 165–206.

Weber, A., Brackmann, A., Both, V., Pavanello, E. P., Anese, RdO. and Thewes, F. R. (2015). Respiratory quotient: Innovative method for monitoring 'Royal Gala' apple storage in a dynamic controlled atmosphere. *Scientia Agricola* 72(1), 28–33.

Whale, S. K. and Singh, Z. (2007). Endogenous ethylene and color development in the skin of 'Pink Lady' apple. *Journal of the American Society for Horticultural Science* 132(1), 20–28.

Whale, S. K., Singh, Z., Behboudian, M. H., Janes, J. and Dhaliwal, S. S. (2008). Fruit quality in 'Cripp's Pink' apple, especially colour, as affected by preharvest sprays of aminoethoxyvinylglycine and ethephon. *Scientia Horticulturae* 115(4), 342–351.

Wills, R. B. H. and Golding, J. B. (2016). *Postharvest: An Introduction to the Physiology and Handling of Fruit and Vegetables*. CABI Publishing, UK. pp. 90–106

Wright, A. H., Delong, J. M., Arul, J. and Prange, R. K. (2015). The trend toward lower oxygen levels during apple (*Malus × domestica* Borkh) storage. *The Journal of Horticultural Science and Biotechnology* 90(1), 1–13.

Yaseen, T., Ricelli, A., Turan, B., Albanese, P. and D'Onghia, A. M. (2015). Ozone for post-harvest treatment of apple fruits. *Phytopathologia Mediterranea* 54(1), 94–103.

Index

www.ingramcontent.com/pod-product-compliance
Ingram Content Group UK Ltd.
Pitfield, Milton Keynes, MK11 3LW, UK
UKHW021008120926
8556IPUK00028B/187